Interstate Eateries

Interstate Eateries, Second Edition
By D.G. Martin

Published by Mann Media Inc.,
Greensboro, N.C.

Interstate Eateries, second edition
© 2008 by D.G. Martin
All rights reserved.

Published by *Our State* magazine, Mann Media Inc.
P.O. Box 4552, Greensboro, N.C. 27404
(800) 948-1409; www.ourstate.com
Printed in the United States by R.R. Donnelley

No part of this book may be used or reproduced in any manner without written permission.

Contents

Preface 6

Interstate 26 9
Interstate 40 17
Interstate 77 41
Interstate 73/74 59
Interstate 85 65
Interstate 95 87

Index of Eateries 98
About the Author 102
Acknowledgements 103

Preface

Home cooking.

It's a phrase that warms our tummies, doesn't it? Barbecue, collards, fried chicken, hot biscuits, real mashed potatoes, banana pudding, sweet tea, and more good old-time North Carolina fare.

Sometimes we get really hungry when we're taking a long trip on one of North Carolina's interstates. And sometimes we crave home cooking, the kind you find in off-the-beaten-path places that only local people know about.

To me, home cooking means more than just good food. It also means eating in a place that makes you feel comfortable — like home. Those eateries are even better when you find a family-owned restaurant that's been around a long time with sons and daughters helping their parents run things. I like it best when waitresses who've been working there forever call me "hon" and keep my glass full of sweet tea without my having to ask. I know I have found my kind of place when the restaurant is full of local people moving from table to table visiting each other, laughing, exchanging news, and maybe even arguing about politics.

When I go to a good home-cooking restaurant, the great food is important, but what I really look for is the welcoming contact with people. These places — those I have found in towns and at crossroads across the state

— give me a way to mingle with people who I would never otherwise meet. It gives me a way to be a small part of another North Carolina community and the lives of the people who live there.

Although our state has lots of these great home-cooking eateries, they can be mighty hard to find when traveling along the interstate. The first *Interstate Eateries* was a compilation of places I found during my travels. In this second edition, I've updated the list with several additional eateries and all the information you need to find them.

It's my hope that your visits to the restaurants in this little book will add the same kind of rich pleasure to your travel as they have to mine.

D.G. Martin
July 2008

Chapter 1

Interstate 26

Running from the mountain border with East Tennessee, through Asheville, Hendersonville, Saluda, and then down to the South Carolina line, Interstate 26 opens the door for travelers to make a quick visit to an eatery in a mountain holler, a small college town, or other mountain hamlets where tourists and local mountain people sit down together for good, solid meals.

Interstate 26

Little Creek Cafe • Mars Hill • Exit 3

"She could flat make biscuits," says Sheila Kay Adams, famous local author and musician, as she remembers the home cooking of the late Edna Boone, who ran Little Creek Cafe for 43 years. The current owner, Bobbie Jo Gillis, with the help of her mother, "Nannie" Littrel, is using years of her own experience running Waffle House restaurants to keep up and build on Edna Boone's legacy.

The biscuits and full breakfast options still draw crowds of locals and, in the summer, visitors from the nearby Wolf Laurel resort. At lunchtime, regulars come in for the "meat-and-two-vegetable plate" specials with drink and dessert and pay less than $7. If you are lucky enough to be there on one of Nannie's chicken and dumplings days, count your blessings!

From I-26
Take Exit 3 to U.S. Highway 23A, and turn left onto U.S. Highway 23A. Little Creek Cafe is just ahead on the right.
11660 U.S. Highway 23, Mars Hill, N.C. 28754
(828) 689-2307
Hours: Monday-Friday, 6 a.m.-2 p.m.; Saturday-Sunday, 7 a.m.-2 p.m.

Wagon Wheel • Mars Hill • Exit 11

For more than 10 years, Wagon Wheel owner Camille Metcalf has been serving breakfast, lunch, and supper to her regular customers in Mars Hill — and to tourists. She gives them a hearty breakfast, lunch specials with everything included for about six dollars, and dinner for well under $10. And for the kids and college students, she always has burgers, fries, and homemade pies.

From I-26
Take Exit 11, and follow N.C. Highway 213 toward Mars Hill for about one mile. Wagon Wheel is on the left.
89 Carl Eller Road, Mars Hill, N.C. 28754
(828) 689-4755
Hours: Monday-Friday, 6 a.m.-8 p.m.; Saturday, 6 a.m.-2 p.m.

Athens Restaurant • Weaverville • **Exit 18**

Bob paced as he waited for Grace outside the Athens Restaurant in Weaverville. They never grew tired of the friendly atmosphere of the place, the welcome they received from the owners and the waitresses, and the good food."

That's how this eatery is described in *At Home in Covington*, a novel by Joan Medlicott, who is a real-life fan of the Athens. Her readers sometimes show up at the Athens to see if it's just like the local restaurant she describes in her series of books about the Ladies of Covington.

"We are on Main Street, but we're not in the middle of town where the tourists gather. So we cater to locals, but we welcome travelers," says one of Athens' owners, Spiros Apostolopoulos.

He brings three generations of restaurant traditions and 25 years of personal experience to Athens. His wife, Elizabeth, and mother, Linda, bake desserts. His partner and brother-in-law, Jimmy Katsigianis, also grew up in a restaurant family. The family has been serving "family food" to folks in Weaverville since 1997. Their real customers enjoy the food just as much as the fictional characters in Medlicott's books.

 *Interstate 26*

From I-26
Take Exit 18. If you're heading southbound, follow U.S. Highway 19-Bus. S. for about one mile (it will become North Main Street). If you're heading northbound, turn right onto Monticello Road; at the first intersection, turn right onto North Main Street. Athens Restaurant is on the left.
247 North Main Street, Weaverville, N.C. 28787
(828) 645-8458
www.theathensrestaurant.com
Hours: Monday-Saturday, 11 a.m.-9 p.m.

Moose Cafe • Asheville • Exit 33

A reader wrote to tell me about the Moose Cafe, located just south of the entrance to the Farmers Market in Asheville, singing the praises of their homemade biscuits, cornbread muffins, sublime apple butter, and iced tea served in mason jars. Not to mention the cafe affords a great view of the mountains and a look at the Biltmore Hotel to boot.

I sampled the mashed potatoes, cabbage, collards, carrots, and pinto beans, and they were farm fresh. Indeed, the cafe's proximity to the Farmers Market ensures a menu that relies on just-picked ingredients.

From I-26
Take Exit 33 for N.C. Highway 191 (Brevard Road). Follow Brevard Road north toward Asheville and the Farmers Market for 1.5 miles. The Moose Cafe is on the right just before you reach the Farmers Market.
800 Brevard Road, Asheville, N.C. 28806
(828) 255-0920
Hours: Daily, 7 a.m.-8 p.m.

Harry's Grill and Piggy's Ice Cream
Hendersonville • Exit 49

"Back in 1980," says owner "Piggy" Thompson, "when my late husband Harry told me he was going to start an ice cream stand here, I told him that I sure was not going to run it. But, when it was time to open up, Harry was still working at his regular job. So, guess what: I came down here and started dipping ice cream, by myself."

Piggy's Ice Cream started by serving more than 20 flavors of Biltmore ice cream, and the business began to grow. After a while, Harry Thompson decided to open a restaurant as well. Although he passed away before it opened, his wife opened it in 1993 and named it after him.

"Piggy" Thompson says she's very proud of the barbecue served at Harry's Grill. It's wood smoked and served with a tomato-based sauce.

From I-26
Take Exit 49 for U.S. Highway 64 West, and head toward Hendersonville for 1 mile. Turn right on Dana Road (becomes Duncan Hill Road). After crossing East 7th Avenue, you'll see Harry's Grill on the right.
102 Duncan Hill Road, Hendersonville, N.C. 28792
(828) 692-1995
Hours: Monday-Saturday, 10:30 a.m.-8 p.m.; during summer, Sunday, 1-9 p.m. (for ice cream only)

Ward's Dairy Bar & Grill • Saluda • Exit 59

For a fine country-cooked breakfast or lunch, amble down Main Street in Saluda, and stop in at Ward's Dairy Bar & Grill, where the specialty is homemade chili. "It's probably the reason why we sell so many hot dogs and hamburgers," former owner Charlie Ward

Interstate 26

would explain.

Recently Ward sold the grill and the adjoining general store and market to Larry and Debra Jackson. Larry Jackson explains it's a challenge to keep up Charlie's great traditions. "People come in and ask for Charlie's sausage. I have to tell them that it's Larry's sausage made with Charlie's recipe."

From I-26

Take Exit 59, and follow the signs toward Saluda, traveling Louisiana Avenue (Ozone Road) for 1 mile to U.S. Highway 176. Turn right on U.S. Highway 176 (Main Street), and then go another half mile to Ward's Grill.

24 Main Street, Saluda, N.C. 28773

(828) 749-2321

Hours: Monday-Saturday, 6 a.m.-3:30 p.m.

Green River Bar-B-Que • Saluda • Exit 59

For almost 25 years, Melanie Talbot has been serving eastern North Carolina-style barbecue to western North Carolinians. And they love it.

Along with a small pork plate for about $10, she will include your choice of three sides, including some options that you won't find other places, like tomato pie, Vidalia onion slaw, sweet potato fries, or corn nuggets with creamed corn in the middle. No wonder it's a popular gathering place for Saluda residents and tourists from all over.

From I-26

Take Exit 59, and follow the signs to Saluda, traveling Louisiana Avenue (Ozone Road) for 1 mile to U.S. Highway 176. Turn right on U.S. Highway 176 (Main Street), and just over the bridge, Green River

will be on your left.
131 U.S. Highway 176, Saluda, N.C. 28773
(828) 749-9892
www.greenriverbbq.com
Hours: Tuesday-Saturday, 11a.m.-8 p.m.; Sunday, noon-3 p.m.

Caro-Mi Dining Room • Tryon • Exit 67

On the porch of the Caro-Mi just outside Tryon, you can wait for a wonderful meal while listening to the rushing sounds of the Pacolet River. Charles Stafford, a former teacher and school administrator, has owned Caro-Mi for 18 years. He's proud that two of North Carolina's leading food experts tout his restaurant — each for different reasons.

Jim Early, author of *The Best Tar Heel Barbecue: Manteo to Murphy*, raves about the skillet-fried chicken livers and mountain trout, along with the vinegar-based shredded cole slaw. Bob Garner, of UNC-TV and *Our State* magazine fame, especially recommends the old-fashioned North Carolina "climate-cured" country ham served here.

You can't go wrong trying either one.

From I-26

Take Exit 67 for N.C. Highway 108, Tryon. Follow N.C. Highway 108 toward Tryon for 2.3 miles to Harmon Field Road. Bear right, and follow Harmon Field Road for 0.7 of a mile to the intersection with U.S. Highway 176. Turn right, and continue for about 2 miles. Caro-Mi is on the left.
1433 Highway 176 North, Tryon, N.C. 28782
(828) 859-5200
Hours: Wednesday-Saturday, 5 p.m.-8 p.m.

Chapter 2

Interstate 40

On Interstate 40, you can travel from California to Wilmington without ever stopping at a stoplight. Following the 420 miles that take you from the Tennessee line to Wilmington, you can visit every region of our state and drive by most of its major cities. You will also have the opportunity to eat barbecue cooked and sauced many different ways, to taste the chicken that helps draw Methodists back to Lake Junaluska every year, and to be amazed at the bountiful all-you-can-eat buffets in eastern North Carolina's rural communities. It's a long drive. So take a break along the way, and enjoy any one of these great places to eat.

Interstate 40

Granny's Chicken Palace • Lake Junaluska • Exit 27

Drop by Granny's for some fried chicken after church on a Sunday, and you may find yourself waiting in line for an hour or so. But not to worry: The local folks and the summer visitors are great people to get to know while you wait.

Don't miss the delicious slaw, the green beans and rice and gravy, and the sweet fritters that make you think you really are back at grandmother's table. Owner Mary Ellen Bowen may not be "Granny," but she has been working at the Chicken Palace for more than 20 years.

From I-40
Take Exit 27 for U.S. Highway 19-23 South toward Waynesville; go about 3.5 miles. Then take Exit 103 and follow directions toward the Lake Junaluska Assembly grounds. Granny's is across from the entrance to the grounds.
1168 Dellwood Road, Lake Junaluska, N.C. 28786
(828) 452-9111
Hours: Monday-Saturday, 4 p.m.-8 p.m.; Sunday, 11:30 a.m.-3 p.m. (Closed during the second week in September and on Tuesdays from November to May.)

Clyde's Restaurant • Waynesville • Exit 27

Clyde's is a favorite memory of novelist Sarah Colton, a native of Asheville who now lives in Paris, where she still thinks about the chocolate pie she ate there when she was growing up. It's still on the menu.

Clyde's is about a 10-mile trek from Interstate 40, but it may be your last chance (or first opportunity) to get North Carolina home cooking on the way to or from the Tennessee line.

From I-40

Take Exit 27 for U.S. Highway 19-23 South toward Waynesville for about 10 miles into Waynesville. Take Exit 98, and turn left onto Hyatt Creek Road. Immediately take another left onto South Main Street.
2107 South Main Street, Waynesville, N.C. 28786
(828) 456-9135
Hours: Tuesday-Sunday, 6 a.m.-9 p.m.

Sherrill's Pioneer Restaurant • Clyde • **Exit 27**

Dean and Lisa Sherrill own and run the Pioneer with the help of family. Always a family affair, the Pioneer was operated in the 1960s and '70s by Dean, his brother, and their mother.

Local folks crowd into the small restaurant and fill its 10 booths and all the counter space for breakfast and lunch. Try the homemade vegetable beef soup — a bargain at $3.25 a bowl. At lunch and dinner, locals love the meat-and-three-vegetable specials that cost a little more than $6.

From I-40

Take Exit 27 for U.S. Highway 19-23-74. Continue for about 1.5 miles. Then bear left, and follow signs to U.S. Highway 19-23 North into Clyde, which becomes Carolina Boulevard. Sherrill's is on the right.
8363 Carolina Boulevard, Clyde, N.C. 28721
(828) 627-9880
Hours: Monday-Saturday, 6 a.m.-9 p.m.

Three Brothers Restaurant
Asheville • **Exit 46B or 53B**

Three Brothers has been a family tradition for more than 50 years. The family actually had four brothers, all of whom eventually were partners, but the sign was

Interstate 40

already up when the fourth brother joined the business. Since money was tight, the brothers left the sign the way it was. Today, it's still run by sons of the original brothers.

The menu has plenty of variety, including steaks, Greek specialties, seafood, and fried chicken, and the sandwiches and hamburgers are tasty and modestly priced.

From I-40
For eastbound, take Exit 46B onto I-240; for westbound, take Exit 53B. Continue on I-240 into Asheville to Exit 4C. Eastbound: Turn right on Haywood Street at the end of the exit ramp; Three Brothers is on the right after the first traffic light. Westbound: Turn left on Montford Avenue, then right on Haywood Street. Three Brothers is on the right.
183 Haywood Street, Asheville, N.C. 28801
(828) 253-4971
Hours: Monday-Thursday, 11 a.m.-9 p.m.; Friday, 11 a.m.-9:30 p.m.; Saturday, 5 p.m.-9:30 p.m.

Coach House Seafood & Steak
Black Mountain • **Exit 64**

Some visitors say the seafood Arthur Pappas serves here is better than what they get on the coast. If the outside doesn't look spiffy, go inside to find out why the locals line up for the generous portions of seafood and home cooking that Pappas has been serving up here for about 15 years.

From I-40
Take Exit 64 onto Broadway Street toward Black Mountain; go 0.5 miles. Turn left on State Street, and go 0.5 miles.
508 West State Street, Black Mountain, N.C. 28711
(828) 669-4223
Hours: Tuesday-Thursday, 11 a.m.-9 p.m.; Friday, 11 a.m.-9:30 p.m.; Saturday, 3 p.m.-9:30 p.m.; Sunday, 11:30 a.m.-9 p.m. (Closes earlier in the winter.)

Perry's BBQ • Black Mountain • Exit 64

In 2001, after many years working with the food service at Montreat College, Jack Spencer, with his wife, Susan, acquired this business from its founder, Perry Cox, and moved it from Swannanoa to downtown Black Mountain.

Perry's isn't fancy, and it's not a big place — only about 10 tables. The barbecue, however, has such a good reputation in the area, that sometimes it gets crowded. Customers come from all over the South and bring a variety of barbecue preferences. "So," says Jack, "we give them lots of different choices for sauces — South Carolina mustard based, eastern-style vinegar, and Lexington tomato. And we've got a good following for our Texas-style beef brisket."

From I-40

Take Exit 64 onto Broadway Street toward Black Mountain; go 0.5 miles. Turn right on East State Street, and go 0.5 miles.
400-C East State Street, Black Mountain, N.C. 28711
(828) 664-1446, www.perrysbbq.com
Hours: Monday-Thursday, 11 a.m.-7:30 p.m.; Friday-Saturday, 11 a.m.-8:30 p.m.

Countryside BBQ • Marion • Exit 86

If you're willing to weave around a few winding roads, there's a good payoff: tasty salads and sandwiches — including a good barbecue sandwich for less than $4, which is a local favorite. There are also meat-and-vegetable plates priced for less than $8. On Friday and Saturday mornings, you'll find a crowd starting the day with a breakfast special — a country-ham biscuit and baked cinnamon apples.

Interstate 40

New owner Rob Noyes, who acquired the restaurant from longtime owners Gary and Lanetta Byrd, brings years of restaurant experience to Countryside.

From I-40
Take Exit 86 for N.C. Highway 226. From the intersection, head north on N.C. Highway 226 toward Marion. Go 1 mile to U.S. Highway 221. Turn right, and go about 0.5 miles. Countryside is on the left.
2070 Rutherford Road (U.S. Highway 221 South), Marion, N.C. 28752
(828) 652-4885
Hours: Monday-Thursday, 11 a.m.-8 p.m.; Friday-Saturday, 7 a.m.-9 p.m.; Sunday, 11 a.m.-3 p.m.

Judge's Riverside Restaurant
Morganton • Exit 103

Some people still call this restaurant Judge's Barbecue. But its current name, Judge's Riverside, reflects a new ambience, and the menu is fancier, too — you can even take it home as a souvenir.

Judge's offers great food, including barbecue sandwiches (both chopped pork and Texas-style beef) and seafood dishes. Worth a side trip, the restaurant sits above one of the few places on the Catawba River where it still flows freely. You can sit on porches overlooking the river. I don't know of any other place this close to one of our interstates where you can get a good meal and enjoy such a nice setting.

From I-40
Take Exit 103, and head north on U.S. Highway 64. Go 1 mile to Fleming Drive. Turn left, and go 1 mile to the intersection with U.S. Highway 70 (West Union Street). Turn left, and go 0.5 miles. Turn right onto Greenlee Ford Road.
128 Greenlee Ford Road, Morganton, N.C. 28655

(828) 433-5798, www.judgesriverside.com
Hours: Sunday-Thursday, 11 a.m.-9 p.m.; Friday-Saturday,
11 a.m.-10 p.m. (Closes an hour earlier in the winter months.)

Hursey's Bar-B-Q • Morganton • Exit 103

You may have thought Hursey's was in Burlington. "Well, you're almost right," says Mike Starnes, owner of Hursey's Bar-B-Q. "I am the only franchisee of Hursey's; I worked with them in Burlington and learned how they cook."

Starnes grew up in Charlotte and opened his business in Morganton in the early 1990s, building a big pit where he cooks the barbecue with wood coals just like Hursey's in Burlington. Barbecue is the specialty, but Starnes says the local crowd loves catfish and chicken, and he is happy to oblige.

If you're hungry for good Burlington-style barbecue, then Hursey's is worth the few extra minutes from the interstate.

From I-40
Take Exit 103 for U.S. Highway 64, Morganton. Head north on U.S. Highway 64. Go 1 mile to Fleming Drive. Turn left, and go 1 mile to U.S. Highway 70 (West Union Street). Turn left; West Union Street becomes Carbon City Road as you go west 1 mile.
300 Carbon City Road, Morganton, N.C. 28655
(828) 437-3001
Hours: Wednesday-Saturday, 11 a.m.-8 p.m; Sunday, noon-3 p.m.

Snack Bar • Hickory • Exit 123B

The Snack Bar serves a variety a solid, home-cooked dishes — mostly meat-and-two-vegetables plates — priced at five or six dollars. There's also a soup, salad, and fruit buffet for $5.20. The soup is hearty and thick with beef and vegetables — easily a meal in itself.

Interstate 40

In 1946, Robert Frye opened the Snack Bar with 11 stools at the counter and two tables. His daughter, Libby, and her husband, the late Eddie Yount, ran and expanded the restaurant. Today, Libby and her son, Brad Yount, carry on the tradition.

Local folks have nicknamed it the Longview Country Club. With the regular customers at their usual tables served by waitresses who know their names, it's no wonder they think of it as their own private club. I found, however, that strangers are welcome, too.

From I-40

Take Exit 123B for U.S. Highway 321-North, toward Lenoir. Follow U.S. Highway 321 for 1.5 miles to the first stoplight. Turn right onto 13th Street SW. Continue for two blocks to 1st Avenue SW; turn left. 1346 1st Avenue SW, Hickory, N.C. 28602
(828) 322-5432
Hours: Monday-Thursday, 6 a.m.-9 p.m.; Friday-Sunday, 6 a.m.-10 p.m.

Keaton's BBQ • Cleveland • Exit 162

People do not just "happen by" Keaton's; they go very intentionally. And they're probably going for the fried barbecued chicken. B.W. Keaton, who founded the restaurant in 1953, came from a family of African-American farmers in the area.

In the early days, locals dropped by to get chicken and to drink a beer. If anybody got rowdy, Keaton knew how to keep things under control: Everybody knew that he kept a shotgun under the counter.

Keaton died in 1989. His niece, Kathleen Murray, runs Keaton's now. Today, the shotgun is gone, and the crowd is family oriented and very orderly.

From I-40
Take Exit 162 for U.S. Highway 64 West. Head west for 1.5 miles to the intersection with Woodleaf Road (Note the Keaton's sign on the corner). Turn left. Woodleaf Road becomes Cool Springs Road as you continue 1.5 miles to Keaton's on the right.
17365 Cool Springs Road, Cleveland, N.C. 27013
(704) 278-1619
Hours: Tuesday-Saturday, 11 a.m.-2 p.m., 5 p.m.-8 p.m.

Miller's Restaurant • Mocksville • Exit 170

On the outside, Miller's Restaurant still looks like the same truck stop that Sheek Miller founded back in 1952. Although the restaurant is a little bit off the beaten path today, it's still a community and family gathering place for folks here. Miller's son, Kip, is the current owner and runs the place like his dad did.

The day I visited, I ordered fried squash, baked apples, and creamed potatoes. The vegetables tasted fresh and delicious. The round, crisp hush puppies were good enough to compete with those at North Carolina's best barbecue restaurants. My waitress kept my iced tea glass full, always with a smile.

From I-40
Take Exit 170 for U.S. Highway 601, Mocksville. Follow U.S. Highway 601 for 1.5 miles. At the stoplight at the intersection with Wilkesboro Street, turn left, and go one block.
710 Wilkesboro Street, Mocksville, N.C. 27028
(336) 751-2621
Hours: Monday-Thursday, 6 a.m.-10 p.m.; Friday, 6 a.m.-11 p.m.; Saturday, 6 a.m.-10 p.m.; Sunday, 7 a.m.-10 p.m.

Interstate 40

Deano's Barbecue • Mocksville • **Exit 170**

Owner Dean Allen has been in the barbecue business since high school. This experience brings good results that have been praised by barbecue guru Jim Early, who says Deano's barbecue is "pound-the-table good." It's definitely worth the short trip into downtown Mockville.

From I-40

Take Exit 170 for U.S. Highway 601, Mocksville. Follow U.S. Highway 601 for 1.5 miles. At the stoplight at the intersection with Wilkesboro Street, turn left, and go 0.7 miles. Turn left at Gaither, and then immediately turn right at North Clement Street.

140 North Clement Street, Mocksville, N.C. 27028
(336) 751-5820, www.deanosbarbecue.com
Hours: Tuesday-Saturday, 11 a.m.-8 p.m.

The Diner • Winston-Salem • **Exit 188**

For my meal at The Diner, I sampled the breaded pork tenderloin along with my scrambled eggs. Owner Steve Eaton Jr. explained that, while the breaded tenderloin is a mainstay, the most popular item is the black-skillet pan gravy, which is served as a separate dish.

Eaton's grandfather, Raymond Eaton, opened The Diner around 1968 in a nearby building. Raymond learned to cook while in the U.S. Navy and at the old Zinzendorf Hotel in downtown Winston-Salem. Later, he bought a small restaurant with nine stools and a counter for about $300. The business expanded, and in 1979, Raymond Eaton sold the business to his son, who sold it to Steve Eaton Jr. in 1996. Steve's grandmother makes some of the desserts — banana pudding, homemade cobbler, and pies.

From I-40
At Exit 188 (the intersection of I-40, Business I-40, and U.S. Highway 421) take U.S. Highway 421 West. Then take the first exit, Exit 239 for Jonestown Road. Turn right, and head north on Jonestown Road. Go 0.5 miles to Country Club Road. Turn right, and go 1 mile; turn left on North Gordon Drive.
108 North Gordon Drive, Winston-Salem, N.C. 27104
(336) 765-9158
Hours: Monday-Friday, 5:30 a.m.-2 p.m.; Saturday, 6 a.m.-noon.

Little Richard's Bar-B-Que
Winston-Salem • **Exit 188**

Richard Berrier, owner and operator of Little Richard's Bar-B-Que, grew up in Davidson County and learned the art of wood-cooking barbecue while working for Leroy McCarn at the Country Kitchen in Midway. While in high school and college, he worked curb service, then the cash register, and later in the kitchen, learning every aspect of the restaurant business — along with the importance of providing warm, homespun service.

When Berrier graduated from Appalachian State University in 1985, McCarn offered him the chance to run the business, which he did until 1990. Then he struck out on his own. In April 1991, Berrier opened Little Richard's.

"I'm the only one still cooking with all wood in Winston-Salem," says Berrier, who cooks only the pig's shoulders, making for a delicious Lexington-style offering.

From I-40
At Exit 188 (the intersection of I-40, Business I-40, and U.S. Highway 421), take U.S. Highway 421 West. Take the first exit, (Exit 239 for Jonestown Road). Turn right, and follow Jonestown Road for 0.5

Interstate 40

miles to Country Club Road. Turn left, and go 0.5 miles.
4885 Country Club Road, Winston-Salem, N.C. 27104
(336) 760-3457
Hours: Monday-Saturday, 11 a.m.-9 p.m.

The Plaza Restaurant • Kernersville • **Exit 203**

When my friends at the *Kernersville News* told me folks line up for the good country cooking at Plaza Restaurant, I decided to give it a try. The restaurant is spiffy clean, and at lunchtime it was completely full.

Undoubtedly, one of the secrets to the Plaza Restaurant's success is variety. Owner Stephen Kroustalis serves a different set of specials every day, always with a big selection of meats and fresh vegetables, at prices ranging from $7 to $9, just the way his uncle Alex did. Stephen, who became manager in 1999, succeeded his uncle as owner in 2007.

Alex had bought the Plaza Restaurant in 1995 from friends whose success with the restaurant he respected. Nevertheless, he gathered a group of experts to make suggestions about what changes he should make. After due deliberation, they told him, "Change nothing but the lock on the door."

Alex followed that suggestion. He attributed his success to four things: "a clean place, good food, reasonable prices, and service. Everything else will take care of itself." Stephen says he follows those same guidelines.

From I-40
Take Exit 203 for N.C. Highway 66-Kernersville. Head north on N.C. Highway 66 for 2 miles. The Plaza Restaurant is on the left in the Plaza 66 Shopping Center.

806 N.C. Highway 66 South, Kernersville, N.C. 27284
(336) 996-7923
Hours: Monday-Thursday, 11 a.m.-8:30 p.m.; Friday, 11 a.m.-9 p.m.;
Saturday, 7 a.m.-9 p.m.

Prissy Polly's Pig-Pickin' Barbecue
Kernersville • **Exit 203**

The judge looked at me with puzzled disdain when he learned that we had omitted this restaurant from an earlier edition. "How could you miss Prissy Polly's?" asked Judge Dickson Phillips, now retired from the U.S. Court of Appeals and the former Dean of the University of North Carolina at Chapel Hill Law School. "It's one of our favorite places — some of the best barbecue I have ever had." Lots of people agree, and they sometimes line up outside, knowing they can choose either Lexington or eastern-style North Carolina barbecue when they get inside.

When Loran Whaley opened up in 1991 and named his new restaurant after his mother Pauline, whose nickname was "Prissy Polly," he served eastern-style barbecue exclusively. But when his sons, Greg and Gary, noticed that many of their customers were covering their plates with a tomato-based sauce, they decided to add a Lexington-style option. "It has worked out well for us. And the customers like the option," Greg says.

A big plate of barbecue, either style, costs a little more than $6. In addition to slaw (eastern-style or Lexington) and the other usual sides, fresh-cooked vegetables are always an option.

From I-40
Take Exit 203 for N.C. Highway 66-Kernersville. Head north on N.C.

Interstate 40

Highway 66 for 2 miles. Prissy Polly's is on the left across from the Plaza 66 Shopping Center.
729 N.C. Highway 66 South, Kernersville, N.C. 27284
(336) 993-5045, www.prissypollys.com
Hours: Monday-Saturday, 11 a.m.-8:30 p.m.

Stamey's Old Fashioned Barbecue
Greensboro • **Exit 29 (old Exit 217)**

I would go to Stamey's just for a bite of the peach cobbler. Flaky, buttery pastry topping over juicy peaches and surrounded by rich, sweet sauce. And it's still less than a dollar-fifty.

But the main reason most folks go to Stamey's is the barbecue, the product of real "cooked by wood fire" and the expertise of three generations of Stameys. It's famously delicious.

From I-40 Bus.
Take Exit 29 (old Exit 217). Go north on High Point Road about 1 mile, noting the signs to the Greensboro Coliseum. Stamey's is on the left directly across from the coliseum.
2206 High Point Road, Greensboro, N.C. 27403
(336) 299-9888, www.stameys.com
Hours: Monday-Saturday, 10 a.m.-9 p.m.

Allen & Son Pit Cooked Bar-B-Q
Chapel Hill • **Exit 266**

If you pass by Allen & Son, you will be ignoring the advice of Bob Garner, author of *North Carolina Barbecue: Flavored by Time*. Garner says the "homemade, chunky, skins-on french fries are reason enough to stop." But he loves the barbecue, too, which he says is "coarsely chopped into meltingly tender chunks, sprinkled through with shreds of deep brown, chewy outside meat." Not finished with his praise, Garner says that Allen & Son serves "one of the tastiest and most authentic versions of Brunswick stew that I've run across." That just sounds mmm-mmmm, good.

From I-40

Take Exit 266, and head north on N.C. Highway 86. Go 1.5 miles, and turn left.

6203 Millhouse Road (just off Airport Road), Chapel Hill, N.C. 27516
(919) 942-7576

Hours: Tuesday-Wednesday, 10 a.m.-5 p.m.; Thursday-Saturday, 10 a.m.-8 p.m.

Margaret's Cantina • Chapel Hill • Exit 266

Margaret's began as a Mexican and Southwestern-style restaurant. It still holds to its roots, but over the years, since owner Margaret Lundy has emphasized the use of fresh local ingredients, there's been a gradual shift toward local cooking styles as well.

It's become a local gathering spot for Chapel Hill folks even though it's a few miles from downtown. Since it's only a minute away from I-40, it gives the traveler an opportunity for a good meal and, for those who love the place, a quick and easy way to spend a few comfortable minutes in Chapel Hill with the locals.

From I-40

Take Exit 266, and head south on N.C. Highway 86 toward Chapel Hill. Go 0.5 miles. Turn left on Weaver Dairy Road. Turn right into Timberlyne Shopping Center. Margaret's is in the center area of the shopping center.

1129 Weaver Dairy Road, Chapel Hill, N.C. 27516

Interstate 40

(919) 942-4745, www.margaretscantina.com
Hours: Monday-Thursday, 11:30 a.m.-2:30 p.m., 5 p.m.-9 p.m.; Friday, 11:30 a.m.-2:30 p.m., 5 p.m.-10 p.m.; Saturday, 5 p.m.-10 p.m.

501 Diner • Chapel Hill • Exit 270

This diner is the breakfast place for many Chapel Hill people. Hamburgers and other diner food make it popular at lunchtime. At supper, the meatloaf plate is their signature dish. And if there are children jumping up and down in a booth or riding the stools, they're probably my grandchildren enjoying one of their favorite places.

From I-40
Take Exit 270, and head south on U.S. Highway 15-501 toward Chapel Hill. Go 2 miles. Do not take Franklin Street exit — turn right on to Ephesus Church Road, and then immediately turn right again onto a service road and into 501 Diner's parking lot.
1500 North Fordham Boulevard, Chapel Hill, N.C. 27514
(919) 933-3505
Hours: Tuesday-Thursday, 7 a.m.-8:30 p.m.; Friday, 7 a.m.-9:30 p.m.; Saturday, 7 a.m.-8:30 p.m.; Sunday, 8:30 a.m.-2 p.m.

Dillard's Bar-B-Q and Seafood • Durham • Exit 276

Some barbecue fans visit Dillard's for its special "South Carolina-style" mustard-based flavoring. But when I worked for a few months at North Carolina Central University, I learned it was also a community gathering place for "soul food" of all types. More important for me, it was a place where you could find out what's going on in Durham today. Just by looking at all the pictures and mementos on the wall, you can learn a lot of history.

Members of the Dillard family have been cooking here

for three generations, and they know what they're doing. Menu options include some Southern home-cooking dishes that are increasingly hard to find on the road — like chitlins and okra.

From I-40
Take Exit 276, and head north on Fayetteville Street toward Durham for 3.5 miles. Dillard's is on the right just before you come to Hillside High School.
3921 Fayetteville Street, Durham, N.C. 27713
(919) 544-1587
Hours: Tuesday-Saturday, 10.30 a.m.-7 p.m.

Neomonde Raleigh Cafe/Market
Raleigh • **Exit 289 or 293**

Neomonde began as a small baking company back in 1977 by the Saleh family, four brothers who grew up in Lebanon. Over time, the bakery grew and added a deli, which transformed into a full-fledged restaurant in 2000. The menu is classic Mediterranean, but it's genuine home cooking so popular in Raleigh that it draws people from downtown and all over town to its off-the-beaten track location near the State Fairgrounds.

From I-40
Eastbound: Take Exit 289 for Wade Avenue. Follow Wade Avenue for 3 miles. Turn right on to I-440, U.S. Highway 1-South. Go 0.6 miles, and take Exit 3. Westbound: Take Exit 293 on to I-440, U.S. Highway 1-North. Go 2.5 miles, and take Exit 3. At end of ramp, turn left on Hillsborough Street. Go 0.3 miles, and turn right onto Beryl Road. Then after crossing the railroad tracks, make an immediate left, and Neomonde is just ahead on the right.
3817 Beryl Road, Raleigh, N.C. 27607

Interstate 40

(919) 828-1628, www.neomonde.com
Hours: Monday-Saurday, 10 a.m.-9 p.m.; Sunday, 10 a.m.-7 p.m.

Pam's Farmhouse Restaurant
Raleigh • **Exit 289 or 293**

Nancy Olson, the world-famous bookseller at Quail Ridge Books in Raleigh, told me about Pam's. "It's one of the best country cooking places, ever," she says. "It's got the best red-eye gravy in the area, and there are always interesting people there." When we finally met there for lunch one day, I found out what she was talking about. The Southern-style vegetables (collards, okra, and corn) offered with my fried chicken were perfectly "home cooked." I loved the banana pudding and wished that I'd had a little more room.

"Pam Medlin has been in the business since she started busing tables at a restaurant our family owned in Henderson," says Pam's mother, Peggy Robinson. That family tradition continues at Pam's. Her brother, Clay Wade, is a cook, and her sister, Tammy Edgerton, is a waitress. Some of the regular customers who eat breakfast and lunch there every day are like family, too.

From I-40
Eastbound: Take Exit 289 for Wade Avenue. Follow Wade Avenue for 3 miles. Turn right on to I-440, U.S. Highway 1 South. Go 3 miles, and take Exit 2B. Westbound: Take Exit 293 on to I-440, U.S. Highway 1 North. Go 2 miles, and take Exit 2B.
From Exit 2B: At end of ramp, turn left on Western Boulevard. Go 0.5 miles. Pam's will be on the left, but you'll need to go 0.2 miles further, make a U-turn at the traffic light at Heather Drive, and come back to Pam's.
5111 Western Boulevard, Raleigh, N.C. 27606

(919) 859-9990
Hours: Monday-Friday, 6 a.m.-2 p.m.; Saturday, 6 a.m.-noon

State Farmers Market Restaurant
Raleigh • **Exit 297**

Although Gypsy Gilliam and her son, Tony, have added some modern features to this historic gathering place, the key to the good eating at the Farmers Market Restaurant is still the incredible source of fresh vegetables from the state Farmers Market, where food growers from all over the region bring their best produce.

But there's more to it. These folks also know how to cook it right. Squash, greens, collards, beans, corn. And don't forget the biscuits or corn bread, iced tea, and friendly service.

From I-40 (I-440)
Take Exit 297, the Lake Wheeler Road/Dorothea Dix/Farmers Market exit. Head north for 0.25 miles, following the signs to the Farmers Market. The restaurant is the building with the big dome.
1240 Farmers Market Drive, Raleigh, N.C. 27603
(919) 755-1550, www.ncsfmr.com
Hours: Monday-Saturday, 6 a.m.-3 p.m.; Sunday, 8 a.m.-3 p.m.

Stephenson's • Willow Spring • **Exit 319**

This is a favorite eating place for popular mystery writer Margaret Maron, who lives nearby and is a distant cousin of the late Paul Stephenson, the man who started the restaurant. It's now managed by his son, Andy Stephenson.

Even if you don't see Margaret Maron when you visit, you will surely see some of the models for the characters in her books among the diverse people who frequent

Interstate 40

the restaurant. And while you're eating your chicken, barbecue, and slaw, don't forget to leave room for the banana pudding.

Stephenson's is probably the only barbecue restaurant in the country that has a commercial nursery next door. Paul's brother and renowned poet, Shelby Stephenson, says Paul started "messing around" with plants not long after opening the restaurant and pretty soon was selling plants. The business grew. Today the nursery business is strictly wholesale, but that won't keep you from checking it out after your meal.

From I-40
Take Exit 319. Head west on N.C. Highway 210 about 0.5 miles. At the intersection with N.C. Highway 50 turn right, and go 1 mile.
12020 N.C. Highway 50 North, Willow Spring, N.C. 27592
(919) 894-4530
Hours: Monday-Saturday, 10 a.m.-8 p.m. (9 p.m. during Daylight Saving Time)

Meadow Village Restaurant
Meadow (Benson) • **Exit 334**

Betty Womble of Sanford tells me that busloads of folks ride over from Sanford to see the famous Christmas lights in Meadow and then stay to eat at Meadow Village. She raves about the seafood and the homemade desserts. "The chocolate pie is to die for; it's delicious. And there's a really nice salad bar."

Meadow Village's owners, Julia Raynor and her son, Timmy, are proud of their low prices — a little more than $6 for the lunch buffet and about 10 dollars for the huge spreads in the evening and on Sunday at noontime.

From I-40

Take Exit 334, and follow N.C. Highway 96 toward Meadow for 0.7 miles. Turn left onto N.C. Highway 50, and go about 100 yards.
7400 N.C. Highway 50 South, Benson, N.C. 27504
(919) 894-5430
Hours: Sunday, Tuesday, Wednesday, 11 a.m.-2:30 p.m.; Thursday-Saturday, 11 a.m.-9 p.m.

The Country Squire
Between Warsaw and Kenansville • **Exit 364**

Maybe the Country Squire is a little too "upscale" to fit in the "home-cooking" category. But so what if it's a little fancy; it's still a gathering place for folks all over Duplin County. My friend Tom Kenan says he makes it a point to stop by for a meal whenever he visits Liberty Hall, the old Kenan family home place, now restored and open to visitors. Owner Iris Lennon, a native of Scotland, has given her restaurant, inn, and surroundings a festive European touch.

From I-40

Take Exit 364, and follow N.C. Highway Business 24 through Warsaw. After 3 miles, turn right on to N.C. Highway Business 24-50 for 4 miles.
748 N.C. Highway Business 24-50, Warsaw, N.C. 28398
(910) 296-1727, www.countrysquireinn.com
Hours: Sunday-Friday, 11:30 a.m.-2 p.m., 5:30 p.m.-9:30 p.m.; Saturday, 5:30 p.m.-9:30 p.m.

Interstate 40

The Country Kitchen Buffet • Wallace • Exit 385

Don't stop at the Country Kitchen Buffet unless you're hungry enough to enjoy its full offering of Southern-style food — all for a set price of about 10 dollars, even less during the week. You'll find fried chicken, seafood, lots of vegetables, and much more, all cooked the old-fashioned way. Local people stop by just for "Miss Ruth's" hand-chopped barbecue and to enjoy the fellowship with Doug and Paulette Jones, who have owned Country Kitchen for more than 13 years.

From I-40
Take Exit 385, and follow N.C. Highway 41 for 3 miles into Wallace. Turn right onto North Norwood Street, and go for 0.5 miles.
607 N. Norwood Street, Wallace, N.C. 28466
(910) 285-8125, www.dougandpaulette.com
Hours: Monday-Saturday, 11 a.m.-8 p.m.; Sunday, 11 a.m.-4 p.m.

Paul's Place • Rocky Point • Exits 408, 414

Paul's Place isn't exactly a home-cooking restaurant, but the hot dogs are legendary. Go all the way with onions, slaw, and his special sauce — a unique red relish that third-generation owner David Paul says is "a holdover from the World War II era when beef was rationed."

David's grandfather started Paul's in 1928. "Back then, it was open 24 hours a day. When my granddad died back in 1939, they had to nail the doors shut to close the place so they could go to the funeral. Before that, they never had any need for locks. It was always open."

From I-40
Westbound: Take Exit 414. Head west on Holly Shelter Road toward Castle Hayne. Go 1 mile to U.S. Highway 117; turn right, and go 3

miles. Eastbound: Take Exit 408, and head west on N.C. Highway 210.
Almost immediately, turn left onto U.S. Highway 117, and go 3.5 miles.
1725 U.S. Highway 117 South, Rocky Point, N.C. 28457
(910) 675-2345
Hours: Sunday-Thursday, 6 a.m.-10 p.m.; Friday-Saturday, 6 a.m.-11 p.m.

Leon's • Wilmington (Ogden) • Exit 420

One of my favorite home-cooking places near Wilmington is just close enough to the end (or beginning) of I-40 to qualify as an "interstate eatery." For many years, Leon Mavrolas and his family treated locals and visitors to delicious and plentiful breakfast, lunch, and supper for modest prices.

My favorite was his creamy seafood chowder. I tried to persuade Leon to give or sell me the recipe. He politely refused. But he's turned it over to new owners, brothers Jimmy and Bobby Pennington, along with the all the other secret recipes and procedures that make Leon's such a special place. Manager Jason Slays, who has worked at Leon's since he was 15 years old, is committed to continuing Leon's tradition of delicious home-style cooking.

From I-40

Take Exit 420, and head east on Gordon Road. Go 2 miles to U.S. Highway 17. Turn left, and go 1 mile. Leon's is on the right just beyond the Ogden Village Shopping Center.
7324 Market Street, Wilmington, N.C. 28411
(910) 686-0228
Hours: Monday-Saturday, 6 a.m.-9 p.m.; Sunday, 7 a.m.-3 p.m.

Chapter 3

Interstate 77

Interstate 77 takes us from the crowded urban areas where Charlotte pushes up against South Carolina all the way to the sparsely populated mountain border with Virginia. Along the way are a couple of genuine home-cooking restaurants in the Charlotte area, a sandwich shop in Davidson, and a family-style option in a historic house. You can sit down where Charles Kuralt critiqued the barbecue, see where local old-time music and food come together, and dine in a small county seat community.

Interstate 77

John's Family Restaurant • Charlotte • Exit 1

Owner John Tsoulos makes this promise on his menu: "No Canned Vegetables." Tsoulos specializes in local, Southern-style food, but it didn't come naturally. He grew up in Greece, left school and went to work at about 10 years old, got a job on a cruise ship, jumped ship in New York in 1970, and made his way to North Carolina.

John's Family Restaurant has been open at this location since 1992. The restaurant is bright, clean, comfortable, and big enough to seat up to 250 people, which is almost enough seating to take care of the lunchtime crowd.

From I-77
Take Exit 1 to Westinghouse Boulevard. From the intersection, head west on Westinghouse. Go 1.5 miles, and you'll see John's Family Restaurant on the right.
2002 Westinghouse Boulevard
Charlotte, N.C. 28273
(704) 588-6613
Hours: Monday-Saturday, 11 a.m.-9 p.m.

Open Kitchen • Charlotte • Exit 9C or 10A

Stephanie Kokenes is the fourth generation of her family to work in the restaurant business in Charlotte. Her father, Alex Kokenes, manages the Open Kitchen. Her grandfather, Steve Kokenes, and his brother opened the restaurant in 1951. And her great-grandfather, Constantine Kokenes, ran the Star Lunch in downtown Charlotte beginning in the early 1900s.

All that tradition comes together at Open Kitchen, which got its name because Steve wanted his customers to feel free to look inside the kitchen to see how the food was being prepared. That tradition still holds.

The food is home-style Italian, even though the Kokeneses originally came from Greece. Sue Brandon, who waited on our group, informed me that the lasagna is the most popular dish. But John Malatras, who worked with the family for many years, says that the pizza is still a big favorite. "They introduced pizza to Charlotte back in 1952," he says.

"And people would come from all over just to try it out. It's still good."

From I-77
Southbound: Take Exit 10A, Morehead Street. Turn right on West Morehead, and go 3 blocks.
Northbound: Take Exit 9C, Wilkinson Boulevard-U.S. Highway 74 West. Go right on Freedom Drive; continue 0.25 miles to West Morehead Street.
1318 West Morehead Street
Charlotte, N.C. 28208
(704) 375-7449
www.theopenkitchen.net
Hours: Monday-Friday, 11 a.m.-10 p.m.; Saturday, 4-10 p.m.; Sunday, 4-9 p.m.

Lupie's Café • Charlotte • Exit 9 or 11

Guadalupe (Lupie) Durand learned how to cook simple foods from Lillie Mae White, the cook at the old Thompson Orphanage in Charlotte, where Lupie went to live when she was 13. "She pretty much cooked everything from scratch," Lupie remembers.

Since Lupie's opened in 1987, the simple, homemade, inexpensive dishes have drawn a diverse and loyal set of

Interstate 77

fans. "I started making chicken and dumplings because it was cheap," she says. "But people like things plain and simple." The simple food and Lupie's welcoming spirit always make for a pleasant mealtime.

From I-77
Southbound: Take Exit 11 for I-277 S/Brookshire Frwy E/NC-16 S. Merge left onto I-277 South. Go 1.8 miles, and bear left to take Exit 2B at US-74 E/NC-27 E/Independence Boulevard.
Northbound: Take Exit 9 for US-74 E/I-277 N/John Belk Frwy. Follow I-277 for 2 miles. Take Exit 2B onto U.S. Hwy-74 E/Independence Blvd.
For both north- and southbound: Follow Independence for 2 miles. Turn right onto Briar Creek Road, and go 0.5 miles. Turn right onto Monroe Road, go 0.5 miles, and Lupie's is on the left.
2718 Monroe Road
Charlotte, N.C. 28205
(704) 374-1232
www.lupiescafe.com
Hours: Monday-Friday, 11 a.m.-10 p.m.; Saturday, noon-10 p.m.

Soda Shop • Davidson • Exit 30

You might think I'm suggesting the Soda Shop just to get you to visit my hometown of Davidson, but this place would be worth the stop even if it weren't right in the middle of one of the nicest towns in the world.

The collection of pennants and sports photos all over the walls lets you know you're in a college town. For me, the experience is extra special because the Soda Shop is almost exactly like the M&M Soda Shop from my childhood. When I order orangeade and an egg-salad sandwich, they're as good as the ones I remember.

The food makes the place great, but what I like best is

that you can stop at any booth and folks will treat you like a long-lost friend.
From I-77
Take Exit 30. From the intersection, head east on Griffith Street. Go 1.5 miles until you dead-end at the Davidson College campus. Turn right on North Main Street and then go 2 blocks.
104 South Main Street
Davidson, N.C. 28036
(704) 896-7743
Hours: Daily, 8 a.m.-8 p.m. Open Friday until 9 p.m.

Isy Bell's Café • Mooresville • Exit 36

Near Mooresville's downtown, Isy Bell's is more than just a little way from the interstate intersections, but it's a home-cooking treasure. I opted for the special plate: four vegetables. I got mashed potatoes with good beef gravy, cabbage, corn, a combination of okra and tomatoes, and a biscuit plus cornbread.

For dessert, the peach cobbler had a wonderful-tasting, bread-like crust that complemented sweet, fresh peaches. And I've heard the strawberry cobbler is even better, but Isy Bell's owner Mike Kabouris had to give me the bad news that he had just run out of that dish. I guess it gives me one more reason to go back to Isy Bell's — not that I needed another one.

From I-77
Take Exit 36 for Mooresville–N.C. Highway 150. Head east, and go 2 miles to West McLelland Avenue (N.C. Highway 152). Turn right on West McLelland, and go 1 mile to South Main Street (also N.C. Highway 152). Turn left, and go 1.3 miles.
1043 North Main Street

Interstate 77

Mooresville, N.C. 28115
(704) 663-6723
Hours: Monday-Saturday, 6 a.m.-9 p.m.

Lancaster's Bar-B-Que & Wings
Mooresville • **Exit 36**

Lancaster's proudly serves eastern-style barbecue right in the middle of Lexington-style barbecue territory. In keeping with that spirit of competition and to celebrate Mooresville's close connections to the stock car racing industry, the restaurant is decorated with racers' uniforms, flags, photos, and full-size racing cars. Lancaster's even has a full-size school bus on the floor, with tables inside for dining. Lots of locals eat here — and plenty of famous drivers, too.

From I-77

Take Exit 36. Head east on N.C. Highway 150. Go 2 miles, crossing U.S. Highway 21, to the intersection with N.C. Highway 152. Turn left onto Rinehardt Road, and go 1.2 miles.

515 Rinehardt Road
Mooresville, N.C. 28115
(704) 663-5807
www.Lancastersbbq.com
Hours: Monday-Tuesday, 11 a.m.-9 p.m.; Wednesday-Saturday, 11 a.m.-10 p.m.

Julia's Talley House Restaurant
Troutman • **Exit 42 or 45**

Julia's Talley House has been around since 1979 when Julia Shumate first opened her restaurant in the former home and office of a beloved family doctor. In fact, Dr.

Talley delivered Shumate, so she honored his memory by using the Talley name. Inside, the place is more like a house than a restaurant.

On my visit, I got a family-style meal served just like an old-time Sunday dinner, with 10 different bowls of food brought to my table. For those not so hungry, there are "cafeteria-style" options for smaller portions of the same good homemade vegetables, meats, and desserts. Paulette Klein, Julia's daughter, says the fried chicken is the house specialty, but the country ham made the meal for me.

From I-77

Northbound: Take Exit 42 for U.S. Highway 21–Troutman. Turn left on U.S. Highway 21, and go 3 miles.

Southbound: Take Exit 45, and follow signs to Troutman. Turn right onto Amity Hill Road, then left onto Murdock Road. Go 2 miles, and turn left onto U.S. Highway 21. Talley House will be within a couple of blocks.

305 North Main Street

Troutman, N.C. 28166

(704) 528-6962

Hours: Monday-Friday, 11 a.m.-2 p.m., 5 p.m.-8:30 p.m.; Saturday, 5 p.m.-8:30 p.m.; Sunday, 11 a.m.-2 p.m.

Carolina Bar-B-Q • Statesville • Exit 49-B

Gene and Linda Medlin have been cooking pork shoulders over hickory wood coals here since about 1985. "Charles Kuralt stopped by here and ate our barbecue," Medlin says, "and you know what he thought about [it]?" Without waiting for an answer, he says, "Not much. Said there wasn't enough fat and gristle in it."

The Medlins have taken Kuralt's criticism to heart, but

Interstate 77

they haven't changed their "refined" way of cooking. Instead, they've posted a notice prominently beside the cash register: "Extra fat and gristle available on request." In addition, they serve up seafood, chicken, salads, and country vegetables.

From I-77
Take Exit 49-B for Statesville–Salisbury Road. Turn right onto Salisbury Road. Continue toward downtown for 2 miles. Carolina Bar-B-Q is at the corner of Salisbury Road and Front Street.
213 Salisbury Road
Statesville, N.C. 28677
(704) 873-5585
Hours: Monday-Thursday, 10:30 a.m.-8:30 p.m.; Friday-Saturday, 10:30 a.m.-9 p.m.

The Cook Shack • Union Grove • **Exit 65**

On Saturday mornings at The Cook Shack, owners Myles and Pal Ireland host a jam session with dozens of fiddlers and banjo and guitar pickers. It's a gathering that's been called "the oldest continuous bluegrass jam session in the world."

If you can't make the jam sessions on Saturdays, try out the great cheeseburgers through late lunchtime any day of the week. Pal cooks a great breakfast, and on cool days she'll serve soup or plates of pintos and cornbread.

Since Myles and Pal have a small grocery store at The Cook Shack, you can find local folks to keep you company at almost any hour of the day. And music lovers will enjoy browsing the store's walls and shelves crowded with bluegrass and country music memorabilia.

Stamey's Old Fashioned Barbecue, Greensboro • I-40 (page 30)

Ellerbe Springs Inn and Restaurant, Ellerbe • I-73/74 (page 60)

Gary's Barbecue, China Grove • I-85 (page 70)

White Swan Bar-B-Q & Fried Chicken, Smithfield • I-95 (page 92)

State Farmers Market Restaurant, Raleigh • I-40 (page 35)

Caro-Mi Dining Room, Tryon • I-26 (page 15)

Lupie's Cafe, Charlotte • I-77 (page 43)

Our Place Cafe, Spencer • I-85 (page 73)

From I-77

Take Exit 65 (N.C. Hwy. 901-Union Grove). Follow N.C. Hwy. 901 for 1.5 miles toward Union Grove. The Cook Shack is on the left, just past the Fiddlers Grove campground and the fire station.
1895 West Memorial Highway (N.C. Hwy. 901)
Union Grove, N.C. 28689
(704) 539-4353
www.fetherbay.com/CookShack
Hours: Monday-Friday, 8 a.m.-6 p.m.; Saturday, 8 a.m.-4 p.m.

Basin Creek Country Store and Restaurant
Elkin • **Exit 83**

There's no "home cooking" here. But for the 25 years since Paul Shumate opened Basin Creek, it's been a gathering place for Elkin residents hungry for chicken wings, great homemade cheeseburgers, good beverages, and the kind of "everybody knows your name" fellowship that we identify with the television program, "Cheers."

From I-77

Take Exit 83 for U.S. Hwy. 21-North. Go 2 miles to Poplar Springs Road. Turn left onto Poplar Springs Road, then left onto North Bridge Street.
2000 North Bridge Street
Elkin, N.C. 28621
(336) 835-5776
Hours: Daily, 11 a.m.-9:30 p.m.

Lantern Restaurant • Dobson • **Exit 93**

Mount Airy and Elkin may be the biggest cities in Surry County, but the smaller town of Dobson is the county seat. At mealtime, the center of social life in

Interstate 77

Dobson is the Lantern Restaurant, where Clinton and Maxine Dockery have been serving breakfast and country-cooking meat-and-vegetable plates since 1972 in the same location.

The original restaurant burned down about 10 years ago. But Clinton Dockery built it back "pretty much the same, because I liked it like it was."

Folks in Dobson like their small town the way it is, too. You will enjoy a quick visit to the downtown and courthouse — after you've sampled the food and hospitality at the Lantern.

From I-77

Take Exit 93 for Dobson. Head east towards Dobson on Zephyr Road (becomes West Kapp Street) for about 3 miles. At the second stoplight, turn left onto Main Street, and Lantern will be on the right.

304 North Main Street

Dobson, N.C. 27017

(336) 356-8461

Hours: Monday-Saturday, 5:30 a.m.-9 p.m.

Chapter 4

Interstate 73/74

North Carolina's newest interstates (I-73 and I-74) are still under development. But they're far enough along for us to begin to cover some of the family-operated restaurants along this stretch of road south of Greensboro. You'll find opportunities to stop at one of our state's most famous barbecue places, a historic family-owned inn, and places where hard-working local folks gather in Randleman, Asheboro, and Candor.

Interstate 73/74

Ellerbe Springs Inn and Restaurant • Ellerbe • Exit 11

Arriving just a little before lunchtime, I order the under-$7 lunch special that features lima beans, corn, carrots, and mashed potatoes. I also order Ellerbe Springs barbecue that the owners tell me is more like Lexington than eastern-style. It has a nice sweet taste that sets it apart from anything else in the state, and I would go back for more.

At dinner, prices are higher, and the certified Angus beef a little thicker. The most popular time is on Sundays for the buffet after church. People crowd in from all around the area, cheerfully paying about 13 dollars for all you can eat plus desserts.

As I wander around the lobby, I spot a picture of the inn from 1910, showing a group of Confederate war veterans holding a reunion there. The restaurant seats more than 150 people, and the inn has 16 rooms for overnight guests. If you're ever close to Ellerbe, drop by the inn and enjoy a century-old institution and a meal to remember.

From I-73/74

Take the exit marked Millstone Road to N.C. Highway 73 West (planned Exit 11). Follow Millstone Road for 2 miles. At the intersection with U.S. Highway 220, turn right, and go 1 mile. The inn is on the left.

2537 North U.S. Highway 220, Ellerbe, N.C. 28338
(800) 248-6467
www.ellerbesprings.com
Hours: Sunday-Thursday, 8 a.m.-3 p.m.; Friday-Saturday, 8 a.m.-9 p.m.

Blake's Restaurant • Candor • **Exit 28**

In 1947, when Colon Blake came back from Army service in France during World War II, he opened Blake's Restaurant in Candor. At first, he bought an old truck stop, fixed it up, and opened a new restaurant. Then, in about 1955, he built a new structure right beside old U.S. Highway 220 where it ran through Candor. The restaurant became a community gathering place and one of the best places to stop on the north-south route that runs right through the middle of North Carolina.

About 25 years ago, when U.S. Highway 220 Bypass was built around Candor, Blake saw the handwriting on the wall. So he had his restaurant building moved the half-mile or so from the center of Candor to its current location, where the bypass and now the new interstate intersects with N.C. Highway 211.

During my last visit, Blake encouraged me to order the chicken pot pie, saying I shouldn't pass it up, but I selected another one of the specials. My fried chicken, butter beans, corn, and mashed potatoes were mighty good, and, after I finished my special plate, Blake told me to try the coconut cream pie. When I said I was full, he said, "No, you have to try it. I'm going to pay for your dessert." That's a hard offer to resist. Made on the premises, the coconut cream pie is as good a dessert as I can ever remember having.

When Blake became seriously ill in 2007, his staff and family rallied to show him that, thanks to his mentoring, they would be able to keep his restaurant going strong, even if he were not there every minute to check on things. And going strong it is. Colon Blake passed away in early 2008.

Interstate 73/74

From I-73/74

Take the N.C. Highway 211–Candor exit (planned Exit 28). Head west on N.C. Highway 211 toward Candor for a few hundred feet, then turn right into the Blake's Restaurant parking lot.
126 North Hillview Street (N.C. Highway 211 East), Candor, N.C. 27229
(910) 974-7503
Hours: Monday-Friday, 6 a.m.-9 p.m.; Saturday-Sunday, 7 a.m.-2 p.m.

Dixie III Restaurant • Asheboro • Exit 56A

Asheboro attorney Ricky Cox makes his way to Dixie III at least three times a week because "their fried shrimp is so good." His law partner, Alan Pugh, says Ricky is crazy to pass by all the other special plates that Dixie III offers — "meat and three vegetables or spaghetti."

Mark Davidson started the Dixie 25 years ago after having grown up working with his dad, former county commissioner Kenyon Davidson, who started the original Dixie Restaurant in Asheboro more that 40 years ago. So Mark comes from an Asheboro restaurant family that takes pride in "full service — meats, seafood, homemade desserts, plus fresh vegetables in season like pinto beans, fried okra, squash, green beans. We've got customers from every walk of life," says Mark. "Suit, ties, or mud on your shoes, it's all okay with us."

From I-73/74

Take the Asheboro–U.S. Highway 64 exit (planned Exit 56A). Head east on U.S. Highway 64. Go 2 miles, and Dixie III is on the left.
715 East Dixie Drive, Asheboro, N.C. 27203
(336) 625-8345
Hours: Monday-Friday, 11 a.m.-9 p.m.

Blue Mist • Asheboro • Exit 56A

The Blue Mist serves a full menu of breakfast items, steaks, sandwiches, seafood, fresh vegetables, and more. But the main attraction is its pit-cooked barbecue, which the Cox family began serving back in 1948.

Current owners, Jeff Clifton and his wife, Robin, bought the Blue Mist in the mid-1990s. Today, they prepare about 1,500 pounds of hams and shoulders each week. The pit-cooked process takes about 18 hours each day, but the result is well worth the effort.

Some people say the way to identify places with the best barbecue is to see how "human" the pig in the establishment logo looks. The more human — that is, dressed up and engaging in human activities — the better the barbecue. If this is true, then Blue Mist's barbecue must be tops. Their pig is standing up, dressed in a tux and tails — ready to go to the debutante ball.

From I-73/74

Take the Asheboro–U.S. Highway 64 exit (planned Exit 56A). Head east on U.S. Highway 64, and go 6 miles. Blue Mist is on the left.

3409 U.S. Highway 64 East, Asheboro, N.C. 27203
(336) 625-3980

Hours: Monday-Wednesday, 5 a.m.-9 p.m.; Thursday-Saturday, 5 a.m.-10 p.m.; Sunday, 6 a.m.-9 p.m.

Note: Blue Mist has two newer locations that are closer to the Interstate: In Asheboro, Exit 58, at 709 South Fayetteville Street; (336) 629-3980. In Randleman, Exit 65, at 319 West Academy Street; (336) 495-3980.

Interstate 73/74

Soprano's Italian Restaurant • Randleman • **Exit 67**

The man responsible for Soprano's and its reputation for solidly good Italian dishes is not Italian at all. Ossamma Hashish hails from the Mediterranean region — Alexandria, Egypt — but he's been in North Carolina for some time, managing a restaurant in Asheboro for about five years before he opened Soprano's in 2000.

You don't have to sing for your super, but you might sing Soprano's praises after a meal. The Italian food is great, and lots of customers order the lunchtime meat-and-two-vegetables special.

From I-73/74

Take the Academy Street-Randleman exit (planned Exit 67).
Head east on West Academy Street for about 0.5 miles.
638 West Academy Street
Randleman, N.C. 27317
(336) 498-4138
Hours: Daily, 11 a.m.-10 p.m. (until 11 p.m. on Friday and Saturday)

Chapter 5

Interstate 85

I nterstate 85 could be considered North Carolina's main street. It joins our state's major urban areas. Because its intersections are so crowded by development, the search for old-time home-cooking places can be a frustrating one. Still, spots to sit down with local people along the way await you. You can sample barbecue not only in Lexington, but also in Gastonia, Charlotte, Rowan County, High Point, Durham, Butner, Henderson, and Norlina. And if barbecue is not your thing, we offer up a smorgasbord of other suggestions for mingling with the locals and eating home cooking.

Interstate 85

Mountain View Restaurant
Kings Mountain • **Exit 8 or 10B**

If you're traveling back and forth to South Carolina, Mountain View is a good place to say "hello" or "good-bye" to small town North Carolina. In the winter, you can get a good view of King's Mountain, where the patriots won an important battle in the American Revolution. Although Mountain View hasn't been in business long, the owner Nick Mantis, former head cook at the Athens Restaurant in Charlotte, and his daughter Georgia, who is the manager, have earned a loyal local clientele for their country cooking, lunch specials, and tasty Greek dishes.

From I-85
Northbound: Take Exit 8, and follow N.C. Highway 161 toward Kings Mountain for 1.5 miles. Turn left at East King Street, and go 0.5 miles.
Southbound: Take Exit 10B, and follow U.S. Highway 74 West for 1.5 miles. Exit onto East King Street/U.S. Highway 74-Business West and follow East King Street for 1.5 miles.
100 West King Street, Kings Mountain, N.C. 28086
(704) 734-1265
Hours: Monday-Saturday, 11 a.m.-9 p.m.; Sunday, 11 a.m.-4 p.m.

R.O.'s Bar-B-Que • Gastonia • **Exit 17**

The special at this Gastonia institution is the slaw burger, a sandwich filled with nothing but cole slaw. But what slaw! A juicy, mayonnaise-based combination with tomatoes and secret spices that somehow make it wonderful. The barbecue-and-slaw combo is also a treat.

Robert O. Black started this restaurant in 1946. His son, Lloyd, still owns and runs the place with the help of Mark Hoffman, Lloyd's nephew. Either of them can tell you about Gastonia history if you go inside and talk to them. But if you

miss the old-time experience of the drive-in, just wait outside, and let the carhops come to you.
From I-85
Take Exit 17 for Gastonia, U.S. Highway 321 South. Head south on U.S. Highway 321 (North Chester Street), and go 1 mile to Airline Avenue. Turn right, and go 1 mile. (Airline becomes West Gaston Avenue.)
1318 West Gaston Avenue, Gastonia, N.C. 28052
(704) 866-8143, www.rosbbq.com
Hours: Monday-Saturday, 9:30 a.m.-10 p.m.

Kyle Fletcher's BBQ & Catering • Gastonia • **Exit 22**

Kyle Fletcher says he got his start cooking when a friend needed barbecue for a wedding reception. Twelve years ago, he leased the building that formerly housed Bland's Barbecue and revamped it for his "charcoal and hickory" method of cooking.

I've heard some people drive by just to enjoy the smells that come from his old-time cooking. As for the "eastern vs. Lexington" controversy, Kyle says, "I don't put nuthin' on my meat — except hickory and charcoal smoke. That way they've got an option to put whatever sauce they want on my barbecue." Lots of local people agree with Kyle, and sometimes you'll find them lined up to enjoy the generous and reasonably priced wood-flavored servings.

From I-85
Take Exit 22 toward Cramerton/Lowell. Turn left onto South Main Street. Go 0.3 miles, and turn left onto Wilkinson Boulevard. Go 0.4 miles. Kyle's is on the left. You will have to make a U-turn at Westover Street to reach it.
4507 Wilkinson Boulevard, Gastonia, N.C. 28056
(704) 824-1956
Hours: Tuesday-Saturday, 11 a.m.-8 p.m.

Interstate 85

Hillbilly's Barbeque & Steaks • Lowell • Exit 23

Hillbilly's Barbeque used to be one of the favorite places for folks who worked at Pharr Yarns in nearby McAdenville. Those Boston butts are still cooking over hickory wood coals inside, and the aroma is worth the price of the meal. The barbecue is a combination of Lexington-style and eastern North Carolina — with a tomato-based sauce that has a touch of vinegar. I found the combination tasty.

From I-85
Take Exit 23 for Lowell-McAdenville, and turn toward Lowell. Hillbilly's is at the next intersection.
720 McAdenville Road, Lowell, N.C. 28098
(704) 824-8838
Hours: Monday-Thursday, 11 a.m.-9 p.m.; Friday-Saturday, 11 a.m.-9:30 p.m.; Sunday, 11 p.m.-3 p.m.

The Old Hickory House Restaurant
Charlotte • **Exit 43 or 45A**

The Old Hickory House Restaurant is a step back in time to when North Tryon Street was the main road between Charlotte and Richmond, Virginia. Members of the Carter family have been running the place for almost 50 years, and it's not much different today from the early days. Longtime owners and brothers Bob and Gene Carter died a few years ago, but their sons, Kevin (Bob's son) and David (Gene's son), are running it just the way their dads did.

I always enjoy watching and smelling the meat cooking on a brick oven inside the restaurant. You'll probably enjoy the chopped red barbecue in heavy tomato sauce. Old Hickory House style is something different, more like what you would find out West.

The restaurant has a Western look, too. Wooden paneling and a rustic look let you know this place has been around a long time. If you want a taste of history, ask Kevin or David. One of them is almost always on board, ready to talk.

From I-85

Southbound: Take Exit 45A for Harris Boulevard, and go almost 1 mile to North Tryon Street (U.S. Highway 29). Turn right, and go 2 miles. Make a U-turn at the stoplight to get back to the restaurant.

Northbound: Take Exit 43 for N.C. Highway 49 and U.S. Highway 29. Go 0.5 miles, and then take N.C. Highway 49/U.S. Highway 29 exit. Bear left onto Mineral Springs Road. Go about 500 yards to North Tryon Street; turn right, and go 0.5 miles.

6538 North Tryon Street, Charlotte, N.C. 28213

(704) 596-8014

Hours: Monday-Saturday, 11 a.m.-9 p.m.

Townhouse II Restaurant • Kannapolis • **Exit 58**

Be ready to eat when you come to the Townhouse II. As one customer, Beth Snead, told me, "You had better be careful when you make your order. If you ask for the 'large' plate and eat it all, you will be so full that you're just going to want to crawl under your desk and die when you get back to work."

When I stopped by, it was the middle of the afternoon. People were coming in, filling the place up. I wondered why they were here at that hour. Janie Hall, the owner, explained that they were eating an early supper. They have to eat early if they want supper with Janie because she closes up at 6 p.m. Why so early? Well, to open at 5 a.m. the next morning, she has to get up at 1 a.m. If you had to get up that early, you would need to go home and get some rest, too.

One caution. When you pull up to Townhouse II, you will see

Interstate 85

a sign beside the restaurant featuring a scantily clad woman. But don't look for any such entertainment inside. That particular sign advertises a lingerie shop in the adjoining shopping area.

From I-85
Take Exit 58 for Concord-Kannapolis. Take the Concord-U.S. Highway 29 South ramp. Continue for 0.5 miles to U.S. Highway 29A (South Main Street). Turn right, and go 2 miles.
1870 South Main Street, Kannapolis, N.C. 28081
(704) 938-8220
Hours: Monday-Friday, 5 a.m.-6 p.m.; Saturday, 5 a.m.-2 p.m.

Gary's Barbecue • China Grove • **Exit 68**

Gary's has been a stopping point on the route between Charlotte and Greensboro for a long time. Gary's is really a museum of times gone by, with a collection of old advertising signs on the walls and an old-time, fully restored Chevrolet Corvette and Ford Thunderbird on display. Don't be misled by the sign on the side of Gary's building that says, "Cokes 5¢." It's just a reminder of the old days; inside, the Cokes cost $1.39, but refills are free.

As I paid my bill, I asked if the restaurant offered a senior citizen discount. "Oh! There is," said the waitress. "It's 10 percent." Then she looked me sweetly in the eye and said, "You know, we're not allowed to ask folks if they are seniors even when they look old enough. Some people take offense. And you — oh, no! I never would have guessed." She knew she had made my day. I went back to the counter, left an extra tip, and walked out the door, holding my gray head high, a little bit giddy, thinking about Corvettes and drive-ins and days long gone.

From I-85

Southbound: Take Exit 68 for China Grove-Rockland; U.S. Highway 29-Bypass South. Go 1.5 miles on U.S. Highway 29-Bypass.
Northbound: Take Exit 68 for China Grove-Rockland; N.C. Highway 152. Turn left, and go 0.5 miles. Turn right onto Madison Road, following signs to U.S. Highway 29 Bypass/U.S. Highway 601 South. Go 0.5 miles.
620 U.S. Highway 29 North, China Grove, N.C. 28023
(704) 857-8314
Hours: Monday-Saturday, 10 a.m.-9:30 p.m.

Porky's Bar-Bq • China Grove • Exit 68

Mike Reid, the owner, cooks with wood, and for those who insist on pit-cooked barbecue, it makes a big difference. According to Elizabeth Cook, editor of the *Salisbury Post*, "It's the only barbecue place I know of around here that offers Carolina-style slaw (red and vinegary) and what I call Virginia slaw (sweet and mayonnaise-based) like I used to get back home in Fredericksburg." You can't find Virginia slaw at many other places around here.

From I-85

Southbound: Take Exit 68 for China Grove-Rockland; U.S. Highway 29-Bypass South. Go about 1 mile on U.S. Highway 29-Bypass to the exit for N.C. Highway 152. At the first stoplight, take a hard left onto Main Street.
Northbound: Take Exit 68 (China Grove-Rockland). Turn left onto N.C. Highway 152, and go 1 mile to the first stoplight. Take a hard left onto Main Street.
1309 North Main Street, China Grove, N.C. 28023
(704) 857-0400
Hours: Daily, 5:30 a.m.-9 p.m.

Interstate 85

The Farmhouse Restaurant • Salisbury • Exit 75

Owners Dan and Denise Schindelholz are working hard to give the folks here the kind of country cooking that makes them happy. And they are succeeding. At lunchtime, the Farmhouse is filled with locals who tank up on sweet iced tea and order the meat-and-two-vegetables special.

Lots of them, I noticed, pick country ham as their "special" meat. The Farmhouse menu lists nearly 20 vegetables, all prepared from scratch. Desserts are homemade and impossible to resist.

From I-85

Take Exit 75 for Jake Alexander Boulevard. Head southeast on Jake Alexander. (Coming from Greensboro, turn left; from Charlotte, turn right.) Go 0.5 miles.

1602 Jake Alexander Boulevard South, Salisbury, N.C. 28146
(704) 633-3276
Hours: Monday-Thursday, 6 a.m.-8 p.m.; Friday-Saturday, 6 a.m.-9 p.m.; Sunday, 7 a.m.-2 p.m.

Richard's Bar-B-Q • Salisbury • Exit 76B

Owner Richard Moore has been hand-chopping barbecue here since 1979 when he became the manager for the former owners. He changed the name when he took over ownership. Barbecue expert Jim Early writes that Richard's barbecue has "a good smoky pork flavor and deep smoke penetration from the slow roasting."

Richard's is a local gathering place, and I like it for the same reason as Walter Turner, historian at the nearby transportation museum in Spencer, who says he enjoys Richard's "comfortable, cozy atmosphere with nice booths and waitresses that keep my iced-tea glass full."

From I-85

Take Exit 76B and follow East Innes Street (U.S. Highway 52) into Salisbury for 1 mile. Turn right onto North Main Street, and go 0.5 miles. Richard's is on the left.
522 North Main Street, Salisbury, N.C. 28144
(704) 636-9561
Hours: Monday-Saturday, 6 a.m.-8 p.m.

Our Place Cafe • Spencer • Exit 79

Even if it weren't just across the street from the state's wonderful transportation museum (a must see for every North Carolinian), Our Place would be worth a stop for a hungry traveler. It's a reminder of the eateries where our parents and grandparents ate.

Owner-manager Darlene Burnside's collections of old-time signs and soft drink bottles will keep you amused while you wait for one of the cafe's famous beefsteak burgers and foot-long hot dogs. Kip Hale, who volunteers at the museum across the street says, "I love the taco salad — the chili makes the difference, and they always crush the corn chips for me. The Brunswick stew is popular, as well as the corned beef-and-Swiss cheese sandwich. This place has the old-timey feeling. The waitresses will do anything for me."

From I-85

Take Exit 79. Follow Andrews Street toward Spencer for 1 mile. Turn left onto North Salisbury Avenue. Go 0.5 mile, and turn right at 5th Street.
111 5th Street, Spencer, N.C. 28159
(704) 636-0036, www.ourplacecafenc.com
Hours: Monday-Thursday, 11 a.m.- 6 p.m.; Friday-Saturday, 11 a.m.-7 p.m.

Interstate 85

Backcountry Barbecue • Linwood • Exit 88

Backcountry Barbecue is the last Lexington-area barbecue stop if you're going south on Interstate 85 — and the first one if you're heading north. First or last, it's a good stop to make.

There's a big woodpile in the back, and pork shoulders are cooked all day over coals with the heat kept constant using a process developed by owner Doug Cook. The barbecue is then kept warm overnight with the help of electric cookers that keep it from being too moist. The fresh-cooked meat is then pulled from the bone and rough chopped just after the customer places an order — all under the supervision of Tina Brogdon, who has been the manager since 1993.

From I-85
Take Exit 88 for Linwood. Head east on Route 47 (Old Linwood Road) toward Linwood. Go 0.5 miles.
4014 Linwood-Southmont Road, Linwood, N.C. 27288
(336) 956-1696
Hours: Monday-Saturday, 6 a.m.-9 p.m.; Sunday, 11 a.m.-9 p.m.

Jimmy's Barbecue • Lexington • Exit 91

Some people will tell you that Jimmy's Barbecue is as good as barbecue restaurants get. Even those who swear by the Lexington Barbecue restaurant agree that Jimmy's is one of the best substitutes.

Besides serving up great barbecue, Jimmy's has some undeniable advantages. It's very close to the interstate — maybe the easiest of all the restaurants to find. Also, it has a varied menu with lots of options for folks who aren't hungry for barbecue. And it's open in the early morning and on Sunday when many other barbecue and country-cooking places are closed. Although Jimmy Harvey died in 2004, his widow, Betty, and their children are running it just the way he taught them.

North Carolina barbecue guru Bob Garner recommends the chicken that Jimmy's serves Thursday through Sunday. "On those days," Garner writes, "you can just about bet that every one of those 125 seats will be full, at least during peak hours."

From I-85

Take Exit 91 for Lexington, N.C. Highway 8. Head southeast on N.C. Highway 8 (Cotton Grove Road). Go 100 yards.
1703 Cotton Grove Road, Lexington, N.C. 27292
(336) 357-2311
Hours: Wednesday-Monday, 6 a.m.-9 p.m.

Whitley's Restaurant • Lexington • Exit 91

Whitley's Restaurant is only about a mile down the road from Jimmy's, but the distance makes a difference. Nearly all of Whitley's customers are locals, and it's a favorite haunt of artist Bob Timberlake, whose gallery is nearby. To sample a homespun atmosphere, stop here. Another reason to drop in is the variety of choices on the menu, with lots of non-barbecue options.

From I-85

Take Exit 91 for Lexington, N.C. Highway 8. Head southeast on N.C. Highway 8 (Cotton Grove Road). Go 1 mile.
3664 N.C. Highway 8, Lexington, N.C. 27292
(336) 357-2364
Hours: Daily, 6 a.m.-9 p.m.

Lexington Barbecue • Lexington • Exit 96

I've eaten here so many times, I think owner Wayne Monk recognizes me when I come in the door. According to *The*

Interstate 85

Charlotte Observer associate editor Jack Betts, Lexington Barbecue is "regarded by many travelers as the Mother Church of North Carolina barbecue." Bob Garner, author of *North Carolina Barbecue: Flavored by Time*, says, "To the faithful, all roads still lead to Lexington Barbecue."

If you have never eaten here, it should be your first stop. I usually get a chopped tray with slaw, hushpuppies, and sweet iced tea. When I leave, I'm six or seven dollars poorer and feel a million dollars better.

From I-85
Take Exit 96 onto U.S. Highway 64, and head west. After 4 miles U.S. Highway 64 merges with Business-85 South. Two miles later you'll see Lexington Barbecue on the right.
10 U.S. Highway 29-70 South, Lexington, N.C. 27295
(336) 249-9814
Hours: Monday-Saturday, 10 a.m.-9:30 p.m.

Captain Tom's Seafood Restaurant
Thomasville • **Exit 103**

Captain Tom's serves up a delicious selection of broiled and fried seafood platters, just like an old-time fish camp, at reasonable prices — even more reasonable at lunchtime. Florence Matthews eats there two or three times a week. When I told her I thought Captain Tom's was pretty good, she looked me in the eye and said severely that it was more than pretty good — "It's a lot better than what they've got at the beach." From the crowd of people in Captain Tom's that day, I would guess that a lot of folks agree.

From I-85
Take Exit 103 for Thomasville-N.C. Highway 109. Head north on N.C. Highway 109 (Randolph Street). Go 0.5 miles.
1037 Randolph Street, Thomasville, N.C. 27360

(336) 476-6563
Hours: Tuesday-Thursday, 11 a.m.-8:30 p.m.; Friday, 11 a.m.-9:30 p.m.; Saturday, 4 p.m.-9:30 p.m.; Sunday, noon-9 p.m.

Tommy's Barbecue • Thomasville • Exit 103

The barbecue in nearby Lexington gets most of the attention, and the barbecue experts often overlook what is going on in nearby Thomasville. The locals, however, swear by the barbecue and other good food that Tommy Everhart has been cooking here since the 1970s.

I loved the friendly service and the reminder that I was eligible for the 10 percent senior's discount. When you wind your way around downtown to get to Tommy's, don't forget to take a look at the famous giant chair as you cross the railroad tracks.

From I-85

Take Exit 103 for Thomasville-N.C. Highway 109. Head north on N.C. Highway 109 (Randolph Street). Go 1.8 miles on Randolph Street, and, after crossing the railroad tracks (at the big chair), turn right onto East Main Street, and go 0.6 miles. Continue to the left as Main Street turns into National Highway, and go 0.2 miles. Tommy's is on the left.
206 National Highway, Thomasville, N.C. 27360
(336) 476-4322
Hours: Monday- Saturday, 6 a.m.-9 p.m.

Pioneer Family Restaurant & Steakhouse
Archdale • Exit 111

After changing ownership a few years ago and then rebuilding after a fire destroyed the building, Pioneer Family Restaurant still carries on a long-standing tradition of providing a full plate of country cooking and a great community gathering place. The menu features charbroiled steaks, seafood, and sandwiches, but almost everybody goes to the 80-item

Interstate 85

buffet bar for an unlimited selection.

My old friend Johnny Bailey lives nearby, and he and his wife go several times a week. "Why so often?" I asked him. "Well, I guess it's because we're gluttons," he said before he could think. "No, don't [say that]. My wife will kill me." I hope she doesn't, but lots of people say the buffet at the Pioneer is to die for.

From I-85

Take Exit 111 for High Point-Archdale. Head north on U.S. Highway 311 (Main Street), and go 1 mile.

10914 North Main Street, Archdale, N.C. 27263

(336) 861-6247

Hours: Sunday-Thursday, 11 a.m.-9 p.m.; Friday-Saturday, 11 a.m.-9:30 p.m.

Bamboo Garden Oriental Restaurant
Archdale • **Exit 111**

Since 1987, owner Nancy Lim is nearly always at the cash register making her regulars and new patrons feel at home. She was born in Cambodia, of Chinese parents, and tells a wonderful story of how she came to this country and settled in North Carolina. Combination platters and family-style dinners are available; most folks, however, choose the buffet for lunch.

From I-85

Take Exit 111 for High Point-Archdale. Head south on U.S. Highway 311 (Main Street), and go 0.5 miles.

10106-C South Main Street

Archdale, N.C. 27263

(336) 434-1888

Hours: Tuesday-Friday, 11:30 a.m.-2 p.m., 5-9 p.m.; Saturday, 5-9 p.m.; Sunday, noon-2:30 p.m., 5-9 p.m.

Kepley's Bar-B-Que • High Point • **Exit 111 or 118**

Is Kepley's worth a side trip into downtown High Point? You will get a definite "yes" from almost everybody who grew up in High Point, where Kepley's is a community institution that brings back many happy memories.

Kepley's is plain and simple, just like it has been since opening in 1948. That might be the secret of its success. Or maybe it's the special vinegar-based, pepper-flavored barbecue that has kept people coming back again and again for more than 60 years.

From I-85

Northbound: Take Exit 111 onto Main Street/U.S. Highway 311 north toward High Point, and go 6 miles. Southbound: Take Exit 118, and follow I-85 Business toward High Point for 7 miles. Exit onto Main Street (U.S. Highway 311), and go into High Point for 3 miles. Kepley's is just before Main Street intersects Lexington Avenue.

1304 North Main Street, High Point, N.C. 27262

(336) 884-1021

Hours: Monday-Saturday, 10 a.m.-8:30 p.m.

Jack's Barbecue • Gibsonville • **Exit 138**

Jack's is a small place and a little bit out of the way. But the folks are so nice, and downtown Gibsonville is a wonderful slice of small-town North Carolina.

The food here is good, too. Folks brag about the special, Jack's Big Boy, which is a giant beef burger with lettuce, tomatoes, onions, mayonnaise, mustard, and melted cheese. It may sound like something you could get at a fast-food place, but it tastes much different, quite wonderful — just like you probably remember the hamburgers in your favorite place back home.

If you're lucky, famous Gibsonville natives like St. Louis

Interstate 85

Rams football player Torry Holt or North Carolina State University women's basketball coach Kay Yow will be back home and dropping by for a Big Boy while you're there.

From I-85

Take Exit 138 for N.C. Highway 61. Go 3 miles to West Main Street. Turn right.

213 West Main Street, Gibsonville, N.C. 27249

(336) 449-6347

Hours: Monday-Friday, 11 a.m.-7 p.m.; Wednesday, 11 a.m.-6:30 p.m.; Saturday, 11 a.m-3 p.m.

Hursey's Pig-Pickin' Bar-B-Q • Burlington • **Exit 143**

I would stop at Hursey's just for the great smell of hickory wood burning. The aroma comes out of its chimneys all day long. Inside, I like to look at the gallery of people — presidents Reagan, Clinton, and Bush; former Senator Jesse Helms; and former Gov. James Hunt — who have told Charlie Hursey that his barbecue is among the best.

Hursey's is home to a big take-out business. There's also a small, comfortable seating area. The menu is limited to barbecue and fried seafood. Hursey's fans, like *The Charlotte Observer* editorial writer Jack Betts, swear it's one of the best barbecue stops in the state. In my book, Hursey's wins the hushpuppy contest with its fresh, round, golden, crispy cakes that are sweet enough to be dessert. In response to the tastes of some newcomers to North Carolina, Ellen Hursey tells me, they have recently added beef and pork ribs to the menu. Guess what. The ribs have become one of the most popular dishes with natives.

From I-85

Take Exit 143, and go north on N.C. Highway 62 (Alamance Road) 1 mile to N.C. Highway 70 (Church Street).

1834 South Church Street, Burlington, N.C. 27215
(336) 226-1694, www.carolinaharvest.com/hurseys
Hours: Monday-Saturday, 11 a.m.-9 p.m.

A&M Grill • Mebane • Exit 157

According to Gene Upchurch of Raleigh, the barbecue is still cooked on a wood fire, the smoke of which drifts into the restaurant and seasons the entire place with a wonderful aroma. Every now and then, a freight train stops on the tracks across from the grill, and the crew runs in for a bag of barbecue sandwiches to go. Meanwhile, "every rail crossing in Mebane is blocked."

The A&M has much more than barbecue, especially during the week. I got the Monday meat loaf special with two vegetables and hushpuppies for less than $5. If you want a little fancier and more expensive meal, the A&M has an upscale branch next door. But I still recommend the original cinder block building for the best bargain eating in the area.

From I-85

Take Exit 157, and go north on Buckhorn Road 0.5 miles to U.S. Highway 70. Turn left, and go 2 miles.

401 East Center Street, Mebane, N.C. 27302
(919) 563-3721
Grill hours: Daily, 6:30 a.m.-8 p.m. (until 8:30 p.m. on Friday)
Bar hours: Monday-Thursday, 5 p.m.-10 p.m.; Friday and Saturday,
5 p.m.-11 p.m.

Riverside Restaurant • Hillsborough • Exit 164

For the past four years, Dorothy and Leon Lea have been building a fan base at their small restaurant "down by the Eno River" in Hillsborough. Country cooking and soul food is their specialty.

Interstate 85

Their biggest fan may be food expert Bob Garner, who wrote in *Our State* magazine (November 2007) that Dorothy "prepares some of the best baked chicken I've ever eaten, and vegetables like boiled cabbage and squash 'n' onions are truly memorable."

From I-85
Take Exit 164, and follow South Churton Street for 0.8 miles. Turn left onto Orange Grove Road, and immediately turn right onto Exchange Park Lane. Go 0.5 miles.
162 Exchange Park Lane, Hillsborough, N.C. 27278
(919) 245-3663
Hours: Monday-Wednesday, 6:30 a.m.-6:30 p.m.; Thursday-Saturday, open until 7 p.m.

Village Diner • Hillsborough • Exit 164

In 1975, Raymond Stansbury and his wife, Ethel, opened a restaurant in the front yard of his family's home about a mile from the center of Hillsborough. Although Raymond died a few years ago, Ethel, along with her daughter Teresa Cox and her family, continues serving Southern country-style cooking and barbecue (now cooked by Teresa's husband, Mike).

Village Diner is the town's oldest continuously operating restaurant. The generous portions (all you can eat) and reasonable prices make it one of Hillsborough's most popular places to gather and eat.

From I-85
Take Exit 164, and follow South Churton Street for 1.5 miles. Turn left onto West King Street, and go 1 mile.
600 West King Street, Hillsborough, N.C. 27278
(919) 732-7032
Hours: Monday- Friday, 6:30 a.m.-8 p.m.

Bullock's • Durham • **Exit 173 or 174B**

Fans of the place will tell you, "Everybody in Durham goes there." It's true; I saw an old friend just a few minutes after the waitress filled my tea glass. Tonia Butler told me, "I came to get some fresh vegetables because I need something healthy." But when she saw my barbecue, fried chicken, Brunswick stew, slaw, hushpuppies, and sweet tea, she laughed and said to the waitress, "Same as he got."

Celebrities come here, too: Shirley Caesar (the gospel singer and minister) and Ken Starr (the Whitewater special prosecutor), to name a few. On the way out, I saw a whole wall of photos of famous people who had eaten at Bullock's. But none could have eaten more happily than I did.

From I-85

Southbound: Take Exit 174B (northbound, you can reach Hillsborough Road from Exit 173), and take the U.S. Highway 70-Business (Hillsborough Road) exit immediately on the right. Turn left on Hillsborough Road, and turn left on LaSalle Street. Go 1 block, and turn right on Quebec Street.

3330 Quebec Street, Durham, N.C. 27705

(919) 383-3211, www.bullocksbbq.com

Hours: Tuesday-Saturday, 11:30 a.m.-8 p.m.

Bob's Barbecue • Creedmoor-Butner • **Exit 191**

At Bob's, you order at the counter and then sit down at one of the long tables full of friendly folks. Harry Coleman, editor of the local newspaper, joined me. He explained to me that the barbecue at Bob's doesn't fit neatly into either of the Lexington or the eastern North Carolina styles. "It's just good, mild barbecue. And the vinegar-based sauce is mild."

But there's no question about it being North Carolina barbecue. It's pork shoulders; it's cooked long and slow; and it has just the

Interstate 85

right flavor — especially for someone who's just about to drive into Virginia from North Carolina and wants to taste our barbecue for the last time.

Owners Paula Ellington and Carla Magnum are the twin granddaughters of Bob Whitt, the original Bob. One of them is almost always on site, making sure everything is clean and bright, that visitors are greeted, and that the barbecue has the same fresh taste that made their family famous.

Don't leave here without tasting the chocolate chess or pecan pies, made fresh daily.

From I-85

Take Exit 191 for N.C. Highway 56 Butner-Creedmoor. Go east toward Creedmoor about 0.5 miles. Bob's is on the left.
1589 N.C. Highway 56 (Butner Creedmoor Road), Creedmoor, N.C. 27522
(919) 528-2081
Hours: Monday-Saturday, 10 a.m.-8 p.m.

Tony's Country Cottage Café • Oxford • Exit 206

Tony's is a favorite of Oxford businessman and former state legislator Stan Fox, who became a fan of Tony Oakes when Tony was coaching football at high schools in Oxford and Henderson. Lots of other football fans followed Tony when he opened his restaurant about five years ago. He rewards them with modestly priced home-cooking specials. "He serves great food," says Fox, "and don't tell anybody, but he's terribly under-priced."

From I-85

Take Exit 206 onto U.S. Highway 158 West toward Oxford, and go for about a hundred yards. Turn right onto Tabbs Creek Road, and follow it for 0.3 miles.
5593 Tabbs Creek Road, Oxford, N.C. 27565
(919) 693-0808
Hours: Sunday- Friday, 11:30 a.m.- 9 p.m.; Saturday, 4 p.m.- 9 p.m.

Nunnery-Freeman Barbecue • Henderson • **Exit 215**

Nunnery-Freeman Barbecue may be one of your last chances to get traditional North Carolina barbecue if you're driving north into Virginia or your first chance to tank up when you return home from up North.

This restaurant has a big fan base in North Carolina, even though it doesn't cook its barbecue in a pit. In fact, Nunnery-Freeman is the reason a lot of North Carolina barbecue restaurants no longer cook over wood; the Nunnery-Freeman folks invented an electric cooker that's sold all over world. If you're missing Gary's Barbecue, which operated nearby, Gary Freeman, son of the inventor, recently merged the two restaurants at the Nunnery-Freeman location.

From I-85

Take Exit 215 for U.S. Highway 158. Northbound: Turn left on U.S. Highway 158 (Norlina Road). Go a few hundred yards. Southbound: Turn left onto Parhman Road, go a few hundred yards, and cross U.S. Highway 158 (Norlina Road). Nunnery-Freeman is on the right.

1816 Norlina Road (also called North Garnett Street and U.S. Highway 158), Henderson, N.C. 27536

(252) 438-4751

Hours: Tuesday-Saturday, 10 a.m.-8:30 p.m.; Sunday, 10 a.m.-8 p.m.

Roadside Cafe (formerly Clem's Place)
Norlina • **Exit 226 or 229**

Don't expect anything fancy here. In fact, some of its biggest fans call this restaurant a "hole in the wall," and they complain it's hard to find because it's concealed by the car wash next door.

What keeps those fans coming back? One of them says that it's because they serve his most favorite Carolina barbecue. But

Interstate 85

it's not just barbecue; it's the "soul food," including chitlins on Thursdays and Fridays that often draw the biggest crowds here.

Clementine Austin owned and ran Clem's Place for many years. In 2007, she handed ownership over to her daughter, Ursula Johnson, who changed the name to Roadside Cafe. While Ursula serves in the U.S. Army, Clem continues to run the restaurant. "Without pay," she says, "13 hours every day." When I ask her what she gets out of this new arrangement, she says, "All the bills go to my daughter. No more do I have to deal with them."

From I-85

Southbound: Take Exit 229, and follow Oine Road for about 3.5 miles into Norlina. Turn left on U.S. Highway 1/158. Roadside is just ahead on the right.

Northbound: Take Exit 226, and follow Ridgeway-Drewry Road towards Ridgeway for about 2.5 miles. Turn left on U.S. Highway 1/158, and go about 2 miles into Norlina. Roadside is on the right.

112 U.S. Highway 1 South, Norlina, N.C. 27563

(252) 456-2407

Hours: Monday-Thursday, 6 a.m.- 7 p.m.; Friday-Saturday, open until 7:30 p.m.

Chapter 6

Interstate 95

I will tell you a secret: You can pull off the interstate at almost any intersection from Lumberton to Roanoke Rapids, drive a few miles in any direction, and find a local family-owned restaurant. A place where the food tastes home-cooked, the staff is friendly, and the regulars are glad to see you. But just in case you don't have time to do the exploring, here are a number of eateries including restaurants owned by members of the Lumbee Tribe, several popular all-you-can-eat buffets, and some highly praised eastern-style barbecue restaurants.

 *Interstate 95*

Linda's Restaurant • Pembroke • Exit 17

Linda's is not only a great place to get a good home-cooked meal; it also gives a visitor a chance to pay a quick visit to the center of life for the largest group of American Indians in the eastern United States — the Lumbee. When I worked at UNC-Pembroke a few years ago, I always liked to go to Linda's Restaurant at lunchtime to catch up on local news. Linda Sheppard serves country cooking on a buffet, and there will always be a short wait. Everybody seems to know one another. The other guests at Linda's won't interrupt your meal, but many will be glad to answer questions about Pembroke or local American Indian life.

From I-95

Take Exit 17. Go west on N.C. Highway 711 for about 10 miles. Linda's is on the left as you come into downtown Pembroke.
408 East Third Street, Pembroke, N.C. 28372
(910) 521-8127
Hours: Monday-Saturday, 6 a.m.-9 p.m.

Sheff's Seafood Restaurant • Pembroke • Exit 17

It's always a treat for me to eat seafood for supper at Sheff's Seafood Restaurant — especially on Tuesdays when Sheff (that is, owner James Sheffield) serves all-you-can-eat spot. People come from all over to partake. Before supper I like to sit in one of the rocking chairs on Sheff's front porch and wait for the trains to come by.

From I-95

Take Exit 17. Go west on N.C. Highway 711 for about 11 miles into Pembroke. You will see Sheff's right after you cross the railroad tracks.
100 East Third Street, Pembroke, N.C. 28372
(910) 521-4667

Hours: Tuesday-Friday, 11 a.m-2:30 p.m.; 5 p.m.-9 p.m.; Saturday,
4 p.m.-9 p.m.

Fuller's Old-Fashion BBQ • Lumberton • Exit 20

About 20 years ago, Fuller Locklear, a Lumbee Indian, and his wife, Delora, opened a small restaurant near their home, serving fried chicken and seafood, Southern vegetables, and barbecue. It grew gradually at first, but, with a new addition, the facility can now seat more than 300 people.

Locklear built a reputation for his collards, which he grew himself. But he wouldn't serve them unless they were fresh and in season. Fuller and Delora have passed away, but their family is continuing the tradition of good food. So, if you want to eat collards, remember to come in the fall or winter.

From I-95

Take Exit 20 for Lumberton/Red Springs. Head west on N.C. Highway 211 toward Red Springs. Go 0.5 miles.

3201 North Roberts Avenue (N.C. Highway 211), Lumberton, N.C. 28360
(910) 738-8694, www.Fullersbbq.com
Hours: Monday-Saturday, 11 a.m.-9 p.m.; Sun., 11 a.m.-4 p.m.

Tarpackers Restaurant • St. Pauls • Exit 31

The name "Tarpackers" comes, of course, from a combination of "Tar Heel" and "Wolfpack," two nicknames of our state's rival public universities. Tarpackers doesn't try to settle the argument about which school is better, but for about 15 years, owners Linwood and Sara Hayes have been celebrating the history and tradition of both schools. The food is simple, light, and reasonably priced. "The ribs we serve on Thursday, Friday, and Saturday evenings have gotten to be a favorite," Linwood says.

Interstate 95

From I-95
Take Exit 31 for N.C. Highway 20–Saint Pauls. Head east on N.C. Highway 20 (Broad Street), and go 1 mile.
201 West Broad Street, St. Pauls, N.C. 28384
(910) 865-1560
Hours: Tuesday-Friday, 11 a.m.-2 p.m.; Thursday-Saturday, 5-9 p.m.

Fuller's Old-Fashion BBQ • Fayetteville • Exit 52

If you missed Fuller's in Lumberton or can't wait to eat until you get there, you're in luck. The late Fuller Locklear's children have opened a branch of Fuller's in Fayetteville with a buffet just like the original. The Fayetteville location serves the same barbecue as the original Lumberton restaurant. It's all cooked on the grounds of the Locklear home place, where Fuller's began more than 20 years ago.

From I-95
Take Exit 52, and follow N.C. Highway 24 west toward Fayetteville for 3 miles. Turn left onto North Eastern Boulevard (I-95-Bus/U.S. Highway 301) and go 0.3 miles.
113 North Eastern Boulevard. Fayetteville, N.C. 28301
(910) 484-5109, www.Fullersbbq.com
Hours: Monday-Saturday, 11 a.m.-9 p.m.; Sunday, 11 a.m.-6 p.m.

Ernie's Buffet • Dunn • Exit 73

Ernie Starling opened his buffet restaurant in this shopping center location about 15 years ago. He had learned the restaurant and barbecue trade by working for about 13 years at the renown Bill's World Famous Barbecue & Chicken in Wilson. "Well, maybe, toward the end, I was teaching them a few things, too," Ernie says, flashing a wry grin that tells me he means it.

Don't miss the big buffet Ernie serves, stocked with first-rate biscuits, fatback, and turnip greens. Seafood is available on Friday and Saturday nights.

From I-95
Take Exit 73 for Dunn–U.S. Highway 421. Head west on U.S. Highway 421 (Cumberland Street) for 1 mile.
1008 West Cumberland Street, Dunn, N.C. 28334
(910) 892-2225
Hours: Monday-Thursday, 11 a.m.-7:30 p.m.; Friday-Saturday, 11 a.m.-3:30 p.m., 4 p.m.-9 p.m.

Town & Country Barbecue & Seafood
Benson • **Exit 79**

Annie McHone has been running the restaurant for almost 20 years, preparing a buffet table of country vegetables and meats. The restaurant is only open on weekends, but all the customers sing the praises of McHone and the food she prepares.

"It's all good," her customers tell me when I press them to reveal their favorites. McHone says she thinks her banana pudding is about as good as anything she makes. But another diner interrupts us and says, "Oh no, that peach cobbler you make is the best around here." In addition to the buffet, McHone has a small but varied menu of seafood, vegetables, steak, barbecued chicken, and other specialties.

From I-95
Take Exit 79 for Benson-N.C. Highway 27. Head west for 0.5 miles to U.S. Highway 301 (Wall Street). Turn left, and go 0.5 miles.
709 South Wall Street (U.S. Highway 301), Benson, N.C. 27504
(919) 894-7252
Hours: Friday, 11 a.m.-8:30 p.m.; Saturday, 3 p.m.-8:30 p.m.; Sunday, 11 a.m.-2:30 p.m.

 *Interstate 95*

Holt Lake Bar-B-Que and Seafood
Smithfield • **Exit 90**

Just down the road from the White Swan, I found Holt Lake Bar-B-Que and Seafood. Brothers Kevin and Terry Barefoot serve wonderful family-style meals almost all day long. You can order from the menu, but, if you have my kind of appetite, you'll want to eat family style, which is available if you have at least four people in your group.

Believe me, it's worth a special trip for the barbecue and fried chicken; and, if you're willing to pay a little more, you can add shrimp and fish to your choices.

From I-95

Take Exit 90. Head north on U.S. Highway 301 (Brightleaf Boulevard). Go 1.5 miles.

3506 U.S. Highway 301 South (Brightleaf Boulevard), Smithfield, N.C. 27577
(919) 934-0148
Hours: Monday, 11 a.m.-2 p.m.; Tuesday-Saturday, 11 a.m.-9 p.m.

White Swan Bar-B-Q & Fried Chicken
Smithfield • **Exit 90**

The White Swan in Smithfield is an old-time barbecue place, just like the roadhouses of days gone by. Lynwood Parker owns and runs the White Swan along with the adjoining motel and an accounting business, as well as being very active in the political life of Johnston County.

Bob Garner brags about the White Swan's barbecue in his book, *North Carolina Barbecue: Flavored By Time*. Barbecue is its specialty, but the restaurant also serves up great fried chicken, Brunswick stew, and ribs.

The restaurant isn't fancy, and it's not big. A lot of its business is done at the counter, with folks coming in to get take-out orders. When I sat down and placed my order of barbecue and slaw, my meal was delivered to the table in less than a minute, along with a helping of some of the best hushpuppies I ever ate.

From I-95

Take Exit 90. Head north on U.S. Highway 301 (Brightleaf Boulevard). Go 2 miles (passing Holt Lake), and the White Swan is on the left.
3198 U.S. Highway 301 South (Brightleaf Boulevard), Smithfield, N.C. 27577
(919) 934-8913
Hours: Monday-Wednesday, 10:30 a.m.-7:30 p.m.; Thursday-Saturday, 10:30 a.m.-8:30 p.m.; Sunday, 10:30 a.m.-5 p.m.

Wilber's Barbecue • Goldsboro • Exit 95

Maybe you think Wilber's Barbecue is too far from the interstate. Goldsboro is almost 30 miles from I-95. You will have to budget an extra 30-plus minutes to get there and the same amount of time to get back. So why am I including Wilber's? Because I would go a long way out of my way to eat with Wilber Shirley.

The food is great — especially the barbecue and fried chicken. The experience of eating at Wilber's should not be denied to any North Carolinian. In *North Carolina Barbecue: Flavored By Time*, Bob Garner says that Wilber's is one of only a handful of restaurants in eastern North Carolina where barbecue is cooked entirely over hardwood coals.

If you ask *The Charlotte Observer* columnist and barbecue expert Jack Betts about North Carolina's best, he says, "If the question is best barbecue, period, then my answer is Wilber Shirley's peppery, chopped barbecue cooked the old way over hardwood coals just behind his one-story, red-brick restaurant on Goldsboro's east side."

Interstate 95

From I-95
Take Exit 95 for Goldsboro. Head east on U.S. Highway 70. Follow the U.S. Highway 70-Bypass through Goldsboro. Wilber's is on the eastern edge of Goldsboro just before N.C. Highway 111 splits off from U.S. Highway 70.
4172 U.S. Highway 70 East, Goldsboro, N.C. 27534
(919) 778-5218, www.wilbersbarbecue.com
Hours: Monday-Saturday, 6 a.m.-9 p.m.; Sunday, 7 a.m.-9 p.m.

Bill's World Famous Barbecue & Chicken
Wilson • Exit 119

Bill's World Famous Barbecue & Chicken restaurant is a part of the Bill Ellis mega-complex that takes up a city block. A catering business, a separate building for take-out food, and a meeting center surround the restaurant.

At the buffet you can get barbecue with fixings, as well as a plate full of country vegetables, fried chicken, and other meats. Next time, I'm going to fast for a day or two before I go back to Bill's, just to be sure I have room for the pudding and cobbler.

From I-95
Take Exit 119 to Interstate 795 South (U.S. Highway 117-264). Follow I-795 S and take Exit 3 (Downing Road). Follow Downing Road toward Wilson for 2 miles. The Bill Ellis complex is at the intersection of Downing Road and Forest Hills Road.
3007 Downing Road, Wilson, N.C. 27893
(252) 237-4372, www.bills-bbq.com
Hours: Tuesday-Sunday, 11 a.m.-8:30 p.m.; Drive-thru open until 9 p.m.

Parker's Barbecue • Wilson • Exit 119

Parker's cooking method — mostly gas-cooked barbecue — causes some controversy: Can it meet the standards of the barbecue purist? I agree with the barbecue experts who simply overlook the gas-cooked controversy and make their judgments based on the results — and you should, too.

Eric Lippard, one of the managers of Parker's, reminded me, "It's still pit-cooked barbecue, and there aren't enough trees in North Carolina to cook it all."

From I-95

Take Exit 119 to I-795 South (U.S. Highway 117-264). Head east for about 5 miles and take Exit 4 to stay on U.S. Highway 264; go about 0.5 miles to the next exit, and take U.S. Highway 301, and head north about 2 miles.
2514 U.S. Highway 301 South, Wilson, N.C. 27893
(252) 237-0972
Hours: Daily, 9 a.m.-8:30 p.m.

Gardner's Barbecue #301 • Rocky Mount • Exit 138

Judging from the long line of folks waiting to eat at Gardner's Barbecue #1 the last time I was there, it's still a great place to visit and eat. Inside, most customers were eating all they wanted, choosing either to go through a buffet line or to be served family style at the table.

As good and plentiful as the food is, there's more to Gardner's than just eating. The lively red-and-white table coverings are just part of the reason it's so bright and cheerful. People smile at each other — at friends and strangers. Somehow, it's more like a church supper than a restaurant.

When the cash register attendants were too busy counting money to talk to me, Gloria Davis, the assistant manager, took me under her wing and answered all my questions. She told me that her boss, Gerry Gardner, brother of former Congressman and Lt. Gov. Jim Gardner, maintains the long family association with this country-cooking institution.

 *Interstate 95*

From I-95
Take Exit 138 for U.S. Highway 64-East Rocky Mount. Go 3 miles to the U.S. Highway 301 Bypass (Wesleyan Boulevard) intersection. Head north on U.S. Highway 301 Bypass for 1 mile.
1331 North Wesleyan Boulevard, Rocky Mount, N.C. 27804
(252) 446-2983, www.gardnerfoods.com
Hours: Sunday-Thursday, 11 a.m.-9 p.m.; Friday-Saturday, 11 a.m.-9:30 p.m.

Ralph's Barbecue • Weldon • Exit 173

I love Ralph's barbecue and Brunswick stew and hushpuppies and banana pudding. But sometimes I like variety and quantity. That's why I timed my visit to Ralph's at suppertime when the buffet line was open. At lunchtime, after 5 p.m., and all day Saturday and Sunday, you can sample the country vegetables, fried chicken, barbecue, and other meats. Ralph and Mason Woodruff started Ralph's back in 1946; today the restaurant is run by Mason's daughter, Kim Amerson.

From I-95
Take Exit 173 for Weldon-Roanoke Rapids. Head east on U.S. Highway 158 (Julian R. Allsbrook Highway). Go 2 blocks.
1400 Julian R. Allsbrook Highway, Weldon, N.C. 27890
(252) 536-2102
Hours: Daily, 9 a.m.-8:30 p.m.

Broadnax Diner • Seaboard • Exit 176

If you want to experience small-town eastern North Carolina the way it used to be, stop by the Broadnax Diner for a meal as I did one Saturday morning. Owners Johnnie and Carolyn Lassiter were working hard behind the counter trying to keep up with the hungry crowd of customers. They fixed me a "big breakfast" with eggs, bacon, sausage, and hash browns while I learned about the prospects for the corn and peanut crops from a local farmer. The Lassiters bought the restaurant about 12 years ago from the family of Seaboard's mayor, Melvin Broadnax. Their only request, according to Carolyn Lassiter, "was that we keep the name, which we have been proud to do."

From I-95

Take Exit 176, and follow N.C. Highway 46 toward Garysburg for 3.5 miles. Turn left onto U.S. Highway 301, and then immediately turn right onto N.C. Highway 186. Go 7.5 miles, and turn right at Park Street in Seaboard.

306 Park Street, Seaboard, N.C. 27876

(252) 589-2292

Hours: Monday-Friday, 6 a.m.-2:30 p.m.; Saturday, 6 a.m.-10:30 a.m.

INDEX OF EATERIES
by *Interstate*

Restaurant, town   **Page No.**

Interstate 26
Athens Restaurant, Weaverville 11
Caro-Mi Dining Room, Tryon 15
Green River Bar-B-Que, Saluda 14
Harry's Grill and Piggy's Ice Cream, Hendersonville 13
Little Creek Cafe, Mars Hill (Little Creek) 10
Moose Cafe, Asheville 12
Wagon Wheel, Mars Hill 10
Ward's Dairy Bar & Grill, Saluda 13

Interstate 40
501 Diner, Chapel Hill 32
Allen & Son Pit Cooked Bar-B-Q, Chapel Hill 30
Clyde's Restaurant, Waynesville 18
Coach House Seafood & Steak, Black Mountain 20
The Country Kitchen Buffet, Wallace 38
Countryside BBQ, Marion 21
The Country Squire, Warsaw-Kenansville 37
Deano's Barbecue, Mocksville 26
Dillard's Bar-B-Q and Seafood, Durham 32
The Diner, Winston-Salem 26
Granny's Chicken Palace, Lake Junaluska 18
Hursey's Bar-B-Q, Morganton 23
Judge's Riverside Restaurant, Morganton 22
Keaton's BBQ, Cleveland 24

Leon's, Wilmington (Ogden) 39
Little Richard's Bar-B-Que, Winston-Salem 27
Margaret's Cantina, Chapel Hill 31
Meadow Village Restaurant, Meadow (Benson) 36
Miller's Restaurant, Mocksville 25
Neomonde Raleigh Cafe/Market, Raleigh 33
Pam's Farmhouse Restaurant, Raleigh 34
Paul's Place, Rocky Point 38
Perry's BBQ, Black Mountain 21
The Plaza Restaurant, Kernersville 28
Prissy Polly's Pig-Pickin' Barbecue, Kernersville 29
Sherrill's Pioneer Restaurant, Clyde 19
Snack Bar, Hickory 23
Stamey's Old Fashioned Barbecue, Greensboro 30
State Farmers Market, Raleigh 35
Stephenson's, Willow Spring 35
Three Brothers Restaurant, Asheville 19

Interstate 77

Basin Creek Country Store and Restaurant, Elkin 57
Carolina Bar-B-Q, Statesville 47
The Cook Shack, Union Grove 48
Isy Bell's Cafe, Mooresville 45
John's Family Restaurant, Charlotte 42
Julia's Talley House Restaurant, Troutman 46
Lancaster's Bar-B-Que & Wings, Mooresville 46
Lantern Restaurant, Dobson 57
Lupie's Cafe, Charlotte 43
Open Kitchen, Charlotte 42
Soda Shop, Davidson 44

Interstate 73/74
Blake's Restaurant, Candor 61
Blue Mist, Asheboro 63
Dixie III Restaurant, Asheboro 62
Ellerbe Springs Inn and Restaurant, Ellerbe 60
Soprano's Italian Restaurant, Randleman 64

Interstate 85
A&M Grill, Mebane 81
Backcountry Barbecue, Linwood 73
Bamboo Garden Oriental Restaurant, Archdale 78
Bob's Barbecue, Creedmoor-Butner 83
Bullock's, Durham 83
Captain Tom's Seafood Restaurant, Thomasville 76
The Farmhouse Restaurant, Salisbury 71
Gary's Barbecue, China Grove 70
Hillbilly's Barbeque & Steaks, Lowell 68
Hursey's Pig-Pickin' Bar-B-Q, Burlington 80
Jack's Barbecue, Gibsonville 79
Jimmy's Barbecue, Lexington 74
Kepley's Bar-B-Que, High Point 78
Kyle Fletcher's BBQ & Catering, Gastonia 67
Lexington Barbecue, Lexington 75
Mountain View Restaurant, Kings Mountain 66
Nunnery-Freeman Barbecue, Henderson 85
The Old Hickory House Restaurant, Charlotte 68
Our Place Cafe, Spencer 73
Pioneer Family Restaurant & Steakhouse, Archdale 77
Porky's Bar-Bq, China Grove 71
Richard's Bar-B-Q, Salisbury 72

Riverside Restaurant, Hillsborough 81
Roadside Cafe, Norlina 85
R.O.'s Bar-B-Que, Gastonia 66
Tommy's Barbecue, Thomasville 77
Tony's Country Cottage Café, Oxford 84
Townhouse II Restaurant, Kannapolis 69
Village Diner, Hillsborough 82
Whitley's Restaurant, Lexington 75

Interstate 95
Bill's World Famous Barbecue & Chicken, Wilson 94
Broadnax Diner, Seaboard.. 96
Ernie's Buffet, Dunn 90
Fuller's Old-Fashion BBQ, Lumberton 89
Fuller's Old-Fashion BBQ, Fayetteville 90
Gardner's Barbecue #301, Rocky Mount 95
Holt Lake Bar-B-Que and Seafood, Smithfield 92
Linda's Restaurant, Pembroke 88
Parker's Barbecue, Wilson 94
Ralph's Barbecue, Weldon 96
Sheff's Seafood Restaurant, Pembroke 88
Tarpackers Restaurant, St. Pauls 89
Town & Country Barbecue & Seafood, Benson........... 91
White Swan Bar-B-Q & Fried Chicken, Smithfield 92
Wilber's Barbecue, Goldsboro 93

About the Author

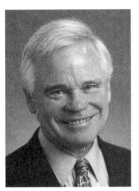

D.G. Martin, who grew up in Davidson, North Carolina, earned a degree in history from Davidson College. After two years in the United States Army Special Forces, D.G. went on to get a law degree from Yale, returning to practice in Charlotte.

D.G. later became vice president of the University of North Carolina System, and he has recently served as an interim vice chancellor at the University of North Carolina at Pembroke and North Carolina Central University, the Carolinas director of the Trust for Public Land, interim executive director of the Triangle Land Conservancy, and interim director of the North Carolina Clean Water Management Trust Fund.

Currently, D.G. writes a weekly column, "One on One," published in about 40 North Carolina newspapers, and he's the host of the UNC-TV "North Carolina Bookwatch" series, which airs Sundays at 5 p.m. He is also a regular contributor to *Our State* magazine and the host of "Who's Talking" on WCHL radio in Chapel Hill.

D.G. and his wife, Harriet, make their home in Chapel Hill, near their children and grandchildren.

Acknowledgements

This book about food and fellowship has had so many helpful cooks that it's impossible to name and thank them all. My wife, Harriet, deserves the most thanks for her patience and for her great editing and writing skills. My son, Grier; daughter, May; and their spouses and children get my appreciation for their cheerful tolerance and loving support.

I acknowledge the valuable help of the many North Carolina barbecue experts, including Bob Garner, Jim Early, Jack Betts, and John Shelton Reed, who are my teachers and guides. Jack and Ruby Hunt taught me so much about how the hospitality that goes with good country cooking brings people together. Also, lunchtime explorations with Bill McCoy, Dick Spangler, and the late Ham Horton inspired this project.

Many people told me about or took me to places I would never have found on my own: Frank Daniels, Walter Turner, David Perry, Norfleet Pruden, Hugh and Brenda Barger, Scott Bigelow, Tom E. Ross, Joe Oxendine, Phil Baddour, Bob Auman, John Curry, Dick Huffman, T. Jerry Williams, Ken Ripley, Thad Woody, Robert and Cynthia Bashford, Betty Kenan, George Couch, Charlie Whitley Jr., Stewart McLeod, Dr. Ben Barker, Grant McRorie, Beth Snead, Al Brand, Stephen Bryant, Charles Blackburn Jr., Jeanne Rhew, Richard Greene, Fountain Odom, Judge Bob Hunter, Libby Lindsay, Eric Millsaps, Tom Terrell, Gene Adcock, Martha McQueen, Sandra Arscott, Ken Ripley, Paul Terry, Dick Hills, Randy Austin, Donna Campbell, and Georgann Eubanks.

The editors and staff at *Our State* have been the greatest partners. My thanks go to Vicky Jarrett, Diane Jakubsen, Louisa J. Dang, Craig Smith, and Meredith Rhoades for all they did to shape this second edition.

Finally, the families who continue to own and operate the restaurants included in this book are my heroes. Thanks to them for giving us these treasured places to gather, meet new friends, and enjoy home cooking.

—*D.G. Martin*